UPDATED AND EXPANDED

THE
STRONGEST
BOY IN
THE WORLD

HOW GENETIC INFORMATION
IS RESHAPING OUR LIVES

Other Titles from Cold Spring Harbor Laboratory Press

THE STRONGEST BOY IN THE WORLD

HOW GENETIC INFORMATION IS RESHAPING OUR LIVES

PHILIP R. REILLY

COLD SPRING HARBOR LABORATORY PRESS
Cold Spring Harbor, New York • www.cshlpress.org

The Strongest Boy in the World
How Genetic Information Is Reshaping Our Lives
Updated and Expanded

Publisher and Acquisition Editor	John Inglis
Development Director	Jan Argentine
Project Coordinator	Mary Cozza
Permissions Coordinator	Maria Fairchild and Carol Brown
Production Editor	Kaaren Hegquist
Desktop Editor	Susan Schaefer
Production Manager	Denise Weiss
Marketing Manager	Ingrid Benirschke
Sales Manager	Elizabeth Powers
Cover Designer	Michael Albano

Front cover artwork: Atlas holding the world on his shoulders. (Image from Creative/Getty Images.)

Library of Congress Cataloging-in-Publication Data

Reilly, Philip, 1947-
  The strongest boy in the world : how genetic information is reshaping our lives / Philip R. Reilly. -- Updated and expanded
      p. cm.
  Includes bibliographical references and index.
  ISBN 978-0-87969-831-7 (hard cover : alk. paper)--ISBN 978-0-87969-943-7 (pbk.: alk. paper)
  1. Human genetics--Popular works. 2. Medical genetics--Popular works. I. Title.
  QH431.R383 2008
  599.93'5--dc22

2008007544

10 9 8 7 6 5 4 3

For a complete catalog of all Cold Spring Harbor Laboratory Press publications, visit our website at www.cshlpress.org.

For Christopher Matthew Tallman Reilly

Contents

Introduction

People are of two minds about science. They are enthralled by its power to discover, explain, and apply. They are also frightened by its power to discover, explain, and apply. It is easy to be enthralled by advances in planetary science and astronomy. Everyone is delighted to see the Mars Rover chugging across the seas of red sand, and we all marvel at the profoundly beautiful photos of supernovas. However exciting, these discoveries concern elements of the universe so far removed from our daily lives that they do not threaten. In sharp contrast, many people are ambivalent about genetics and the biological sciences. Almost everyone eagerly supports research to develop new drugs against cancer or to perfect a vaccine that immunizes against transmission of HIV. But genetic research that investigates our origins, explores the genetic boundaries of race, compares the human DNA sequence to that of chimpanzees to identify gene variants that correlate with intelligence, explores the possibility of extending the human life, or investigates how we might reengineer our species evokes a decidedly mixed response. I wrote these chapters in part to ameliorate the unease about molecular genetics. I think that one way to do this is to explore some of the issues concerning use of new knowledge that trouble laymen and scientists alike.

I organized the first 5 of my 21 essays in a section entitled "Humanity." I did so because the topics raise issues about how we perceive ourselves. The first chapter in this book is about a boy in Germany who was born with a mutation in a gene that results in his having extraordinarily large muscles. Such mutations have been found in other animals. In Europe there are several strains of "double muscled" cattle that result from mutations in the same gene that is mutated in the little boy. Medical reports about this child stimulated me to think about the impact that genetic testing and (somewhat further in the future) genetic engineering might have on the recruitment, training, and public response to elite athletes. Could it be that the current controversies about steroid use will a decade or two hence be regarded as quaint? Will we harness genetic knowledge in ways that confer far greater advantage than any designer drugs ever could? I figured that a story about the future of athletics should entice many readers who might otherwise pass on this book!

Few topics are more gripping than the story of human origins (Chapter 2). The discoveries of physical anthropologists and molecular geneticists who specialize in studying ancient DNA suggest the tantalizing possibility that as many as *four* different *species* of early humans may have occupied the planet 30,000 years ago. What happened to the other three? Did they die off of natural causes or did our ancestors push them into extinction? My chapter on our ancestors attempts to provide an overview of an immensely complicated story that is in a constant state of revision. If I get some readers interested in digging deeper, I will be more than satisfied. On the other hand, I know that no person or culture easily modifies an origin story in light of new knowledge. Much as Darwin's theory of evolution through natural selection shook mid-19th century theology, advances in genetic science today challenge deeply held religious beliefs. Those who believe firmly in an origin story that begins a few thousand years ago in the Garden of Eden are not likely to embrace this chapter.

They may also wish to skip the next chapter on race. Interestingly, both physical evidence and logic suggest that racial divergence within the human species is a recent phenomenon that manifested only in the most recent 40,000 years of our history. In this chapter, I try to summarize the origins of our modern thinking about race. More importantly, I make the point that genetics has proven that race is only skin deep. The biological differences between races are minuscule compared to the similarities. If they share the same versions of a handful of genes, any person on the planet can be a kidney donor to any other person. Now, that is similarity!

Few fields excite me more than the current work in the genetics of longevity. Studies on organisms ranging from the lowly worm, *C. elegans*, to our cousin, the mouse, *Mus musculus*, strongly indicate that a few genes may have very important effects on aging. On-going experiments with worms, flies, and mice demonstrate that it is possible to breed long-lived strains of animals. Imagine the consequences of identifying longevity genes in people and, more importantly, learning how to improve their function! No one can deny that society's interest in effective gene-based approaches to longevity will be overwhelming. But what kinds of social problems arise when only some groups have access to life-extending technologies?

Work on detecting gene variants that have a measurable influence on intelligence (which I define here only as a set of scores derived from taking a set of conventional tests) is proceeding slowly. It is likely that the brain is so complex that no one gene can account for even one percent of the expression of such a complex trait. On the other hand, scientists have shown that a mutation in a single gene called FOXP2 can profoundly limit the ability of a person to use language. Perhaps there are key genes after all. Imagine intelligence tests being supplanted by gene tests, with the results used to guide expectations or manage education! In the not too distant future we will be able to acquire vast amounts of genetic data about people at very low cost.

Most of us are fascinated by the steady progress of genetic and related research to provide an ever more sophisticated understanding of the origin of disease and to suggest new paths for amelioration and cure. For the most part, our great successes thus far involve the elucidation of disorders caused by errors in a single gene. My stories about the train of discoveries concerning Charcot-Marie-Tooth disease, a neurological disorder involving the distal muscles, and Huntington's disease, a late-onset fatal disorder of the brain, illustrate our scientific power and hold out hope for cures once thought to be impossible to imagine. It is not morally problematic to support such research.

However, when we consider a condition such as congenital deafness that is quite common and largely genetic in origin, we confront much harder questions about the role of genetics in medicine and society. The most fundamental question is to decide whether or not deafness is a disease. Many deaf adults do not accept that label; many hearing parents of infants newly diagnosed as deaf do. Which is it? Must we decide? Do new technologies such as screening of infants to ascertain whether their deafness is due to genetic mutations in the parents' germ cells, and cochlear implants to pull deaf children into the world of the hearing, herald a slow, inevitable decline in the size of the Deaf Community? It is quite likely that in technologically advanced nations, there will be far fewer deaf adults five decades hence than there are today. Is a culture disappearing?

My story about an uncommon chromosomal disorder called San Luis valley syndrome raises a different set of moral dilemmas. The judicious use of public education programs and selective genetic testing to identify adults at risk for parenting affected kids could lead to the sharp reduction in the birth of children with this severely debilitating congenital disorder. Is this a proper public health activity? Such a program—even if voluntary in nature—raises deep questions about the morality of selective abortion and the definition of what constitutes a serious disease. Yet, we may be nearing the day when DNA analysis of the fetus early in pregnancy is routine and selective abortion is far more common than it is today. Does the right to privacy include the right to terminate a pregnancy for any reason?

The complex, gripping tale about experimental use of gene therapy to cure a fatal disorder called severe combined immune deficiency forces us to anticipate a future in which parents might be able to use genetic technologies not only to cure disease, but also to enhance wellness. Gene therapy raises a host of issues ranging from the ethics of human experimentation to the allocation of scarce resources. If the "haves" use genetic engineering to enhance their children and the "have nots" do not, will the great divides in our society someday be much deeper?

Genetics is reshaping nearly every aspect of our world. How best can I make this case? I thought the chapters on dogs, cats, mice, and corn and rice would make the point. I am particularly intrigued by the idea that a few decades hence scientists will develop new methods to select for traits that we know as intelligence in animals. I know you already think your dog (or cat) is smart, but what will it mean if in 20 or 30 years you can acquire a dog (or cat) that is really, really smart? Another fascinating question is to ponder whether reproductive cloning—currently considered morally repugnant for use in humans by the vast majority of people—will someday be tolerated in part because cloning a favorite pet had become a relatively common activity in the upper middle class.

My essay on mice takes a different tack. These little critters have done so much for humans over the last century or so that I think they deserve an accolade. It turns out that the mouse genome is so similar to the human genome that we can learn profoundly important facts about ourselves from them. The latest, mind-blowing, example of this is the creation of the first mouse strain in the world to carry a whole human chromosome in the nucleus of its cells. The human chromosome is number 21. Adding it to early mouse embryos affects the mouse with a murine version of Down syndrome, one of the most common chromosomal disorders in humans.

By studying the embryological development of these animals, we are going to rapidly learn a great deal about what goes wrong in Down syndrome. Maybe this will generate new approaches to neutralizing some of the harms it causes.

The rise of corn from a humble little plant with perhaps 4–8 kernels in the highlands of Mexico to the thousands of domesticated strains that grow throughout the world today is a story of human ingenuity and the power of observation. It was peasant farmers who over several thousand years created the corn we know. But, since the 1930s, scientists have put the process of artificial selection in maize into overdrive. Genetically modified corn that today can grow in extremely arid conditions may save millions of lives in Africa. Corn with genes built in to resist certain pests will sharply reduce the need for thousands of tons of insecticides in the environment. These are some of the benefits, but what are the risks? The chapter on corn examines the debate over the use of genetically modified food.

Rice is the most common food consumed by humans. It keeps about one-fifth of the people in the world alive. But geneticists are doing a lot more than developing new strains that grow well in harsh climates. Vitamin and micronutrient deficiencies are a major cause of death and disability among children in underdeveloped nations. Progress to develop a strain of rice rich in vitamin A has been steady. Thanks to the work of a handful of dedicated scientists, consumption of vitamin-enriched rice may someday soon prevent hundreds of thousands of cases of childhood blindness each year!

In the final section of the book, I consider the influence of genes on society even more broadly. The examples I offer about the impact of DNA analysis on the ever-changing historical record may seem at first glance to address relatively minor questions. But the question of whether Thomas Jefferson cohabited for years with a slave woman offers a deep insight into the behavior of a key historical figure and may also suggest something broader about race relations in 18th-century colonial America. The "gossip" questions that DNA testing has sought to address should not overshadow deeper questions that someday may be answered with similar technologies—such as the origin and spread of infectious diseases, a story that profoundly shaped the colonization of the New World.

In no quarter of our society has the impact of genetics been more profound than in the criminal justice system. As I write these words, there is a bill before Congress (already approved in the Senate) to permit federal authorities to take a DNA sample following the arrest or detention of any person, a proposal that greatly extends the current reach of DNA felon data banking laws that in most states apply only to persons convicted of certain crimes of violence. Will we soon inhabit a world in which at birth every citizen gives up a DNA sample to the state to analyze and place in a registry? Is this the ultimate way to curtail serial criminals? Is it the only way to operate DNA data banks that do not discriminate against minorities?

Advances in molecular biology and genetic engineering have deeply influenced our linguistic and our artistic expression. Nearly every day I read the newspaper, I find that at least one writer (and, often, several) has used a genetic metaphor to explain someone's ability, interests, or drive. One reads of the "shopping gene," the "collecting gene," the "bridge playing gene," the "car designing gene." The list goes on and on. Cloning is a technology that writers seem to find immensely attractive. In media circles it is now fashionable to pay a compliment to someone by noting that he or she should be cloned. It is fun to collect these quotes and show them off, but is there a deeper issue here? Do trends in our language reflect a growing belief in genetic determinism? Is the ever-shifting nurture–nature debate swinging toward nature? Many modern artists have chosen to create in response to molecular biology. Although not uniformly so, the majority of the works present a brooding, worrisome response to the science. Is this in part due to a misunderstanding? Or should we heed the cautionary message?

During the first half of the 20th century, many nations implemented eugenics programs—efforts to prevent those thought to be at unusually high risk for having children with mental or physical disabilities from reproducing. State-supported eugenics reached its apotheosis in Nazi Germany in the late 1930s, when the state ordered the sterilizations of hundreds of thousands of persons as a warm-up for the mass murder of millions more. A half-century later, the threat of state-sponsored sterilizations seems remote, but eugenic ideas remain. In our era, eugenics is scientifically based, technologically empowered, and individually driven. Most couples who learn that they face a one-in-four risk for bearing a child with a severe genetic disorder want to avoid that outcome. During the last 20 years, prenatal diagnosis and selective abortion provided a (admittedly unpalatable) solution. Today, preimplantation genetic diagnosis (PGD), the topic I take up in Chapter 19, offers a better option. Couples can use in vitro fertilization to create several embryos, use genetic testing to determine which ones are not affected with the disease in question, and implant only those that won the genetic lottery. But PGD poses an ethical dilemma. Prospective parents can now choose to give birth to children who will not have a particular disorder; in the future, they will be able to give birth to children who have certain "preferred" biological characteristics. This era has begun. During the last decade, a small but growing number of parents have chosen to implant and carry children expressly so that after birth they can be bone marrow donors for existing siblings with severe disorders. These have been dubbed "savior children." Some people think it is wrong to conceive a child for such a purpose, but most find it an acceptable act of parental love. However, there are tougher questions looming in the future. In 10 years, will some couples be using PGD to select for genetic traits that enhance the likelihood that a child will be smarter, more athletic, or more musical? Should that be permitted?

Over the last 25 years, no area of science has engaged the public more than stem cell biology. The public debate over the use of cells derived from frozen human embryos crystallizes the tension that I articulated at the top of this introduction. Many people unabashedly support research involving human embryos, rightfully seeing it as offering the best option to make major breakthroughs in treating a number of common, incurable disorders such as Parkinson's disease. Many more, despite their desire to conquer disease, abhor the thought of sacrificing a single human embryo to that goal. Many assert that the debate over the moral status of frozen human embryos is the latest battle in the abortion wars. I think it is that *and* something more. The likelihood that we will be able to isolate stem cells and then manipulate them to become heart muscle cells, pancreatic exocrine cells, brain cells (or any other of the more than 300 cell types in the human body), grow these specialized cells, and then use them to replace their worn-out counterparts is a powerful statement about the future of medicine. It casts biomedical researchers in the role of the sorcerer's apprentice. Will we properly use the powerful tools we are building? The debate in the United States over the moral status of early human embryos and their use in research will only marginally limit the progress of research. The real moral issue is how we will use cell therapy when it has matured.

In expanding and updating the paperback edition of this book, I had no choice but to add a new chapter, which I titled "Personal Genomics." In the last two years, new technologies have become available that permit researchers to conduct much more powerful studies of the influence that genes have on common disorders. In this new chapter, I discuss the advent of whole genome association studies—a research strategy that is keyed to asking which of the thousands of natural variations (called single nucleotide polymorphisms or SNPs) that are distributed across the human genome correlate with a predisposition to disease. In 2007, whole genome association studies discovered gene variants that are associated with increased risk for multiple sclerosis, diabetes, heart disease, macular degeneration, rheumatoid arthritis, and amyotrophic lateral sclerosis (ALS or Lou Gehrig's Disease), to name just a few. The technologies that permit this research are also the foundation for an emerging industry—direct to consumer genetic testing. For a cost of between $1000 and $2000, one can send a DNA sample (obtained by cheek swab) to one of a rapidly growing number of companies and receive an analysis of many thousands of the variations in your DNA, including those that may pose increased risk for certain diseases. Thus far, most of the correlations are of little clinical value, but I think some (such as the genetic risk for macular degeneration) are important to know about, and more will come. Some critics have called this new enterprise "recreational genetics," which may be apt. The emergence of these companies is raising challenging questions about an individual's right to obtain the data versus the government's right to regulate the industry. We may soon face the profound question of how much we want to know about our future health risks and what criteria we should use in deciding that question.

In an essay written as part of a Festschrift in honor of the contribution made by the Albert Lasker Medical Foundation over the last 60 years to the support of biomedical research, the Nobel Laureate, Joshua Lederberg, a founding father of the field, reluctantly addressed the future of molecular biology. He reflected that, "Molecular biotechnologies ha(ve) reached such a state of potential power that there might be no limit to the possible."[1] He went on to assert that humans do not await the future; they make it. He then mused as to whether by the end of this century we might have sequenced the genome of every species on the planet, creating an ultimate repository of biological knowledge. Leaping even further, Lederberg wondered if 200 years from now we might be able to create "more behavioral enhancements than anyone could dare think about"[1] now. A century or two is a wink of the eye on the timescale that measures the rise and fall of a species. If technology will confer such profound powers so soon, then we do well to anticipate their use. I hope this volume is an effective and entertaining way to do that.

[1]Lederberg J. 2005. Metaphysical games: An imaginary lecture on crafting earth's biological future. *J. Am. Med. Assoc.* **294:** 1415–1417.

PART

1

HUMANITY

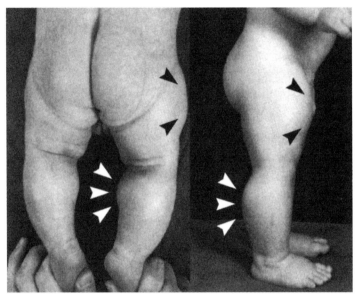

Photograph of the boy with the mutation that causes unusually large muscles. The arrows highlight their unusual bulk. (Reprinted, with permission, from Schuelke et al. 2004 [©Massachusetts Medical Society. All rights reserved.].)

1

The Strongest Boy in the World

I can hold a pair of 3-kilogram (6.6 pounds) dumbbells with my arms extended for 15 seconds or so before I start to feel the strain. My 83 pound, 12-year-old son, who is small but wiry, cannot keep them at the horizontal for much more than 10 seconds. But there is a little boy in Germany who at the age of 4 1/2 easily eclipsed my son's strength and may in a few years match my own. There is a pretty good chance that for his age he is the strongest boy in the world.

This little guy was born after an uneventful pregnancy to a healthy woman who was for a time a professional athlete. The medical reports about the child include no information about the father. The boy's birth weight was normal, as was his newborn physical exam. However, within hours of birth he was noted to have myoclonus—a sudden, jerky contraction of a group of muscles—in response to even a mild stimulus. This led the doctors to keep him in the hospital for evaluation. Closer physical examination, complemented by ultrasound studies, showed that he had unusually large muscles. The myoclonus improved, and because all other aspects of his evaluation were normal, he was sent home. Pediatricians have been closely following his growth, especially his muscle development, ever since.

The boy's myoclonus subsided over a few months, but his muscle mass has continued to be far above normal on every examination. Fortunately, his heart (which, although a muscle, is composed of different tissue than are the muscles in the arms and legs) is normal in both size and function. The little boy continues to have astounding strength for his age. One medical report about him disclosed that several of the boy's maternal relatives are unusually strong. His grandfather was a construction worker who was said to be able to unload large curbstones from a truck, a feat that no one else on the crew could perform. His mother is obviously muscular, although not as dramatically so as her son.

Because of the child's highly unusual development, his mother permitted doctors to perform a number of studies, including genetic testing.

Ultrasound analysis showed that the boy's quadriceps (the large extensor muscle in the front of the thigh that is divided into four compartments) muscle was more than *six* standard deviations above the mean for his age! That is, quite literally, fewer than 1 in 1,000,000 boys would be expected to have muscles this big! In addition, his subcutaneous fat was nearly three standard deviations below the mean. The little boy seems to be on a developmental trajectory that will give him the well-muscled body that many young men dream about—powerful muscles and essentially no fat—but never attain.

The boy's muscles reminded the doctors who examined him of similar patterns of increased muscling that has been found in mice and cattle. Cattle breeders in Europe have long been interested in the occasional birth of animals in several breeds that have unusual muscle bulk and sharply reduced body fat, a phenotype of obvious commercial interest. Efforts to find and clone the causative gene began in earnest in the mid-1990s. In 1997 researchers on the Faculty of Veterinary Medicine at the University of Liege in Belgium, studying a stretch of bovine chromosome two that was known to contain the gene associated with heavy muscling, demonstrated that in Belgian Blue cattle, the heavy muscling was due to a recessive disorder involving a single gene. They found that the unusually muscled cattle had small *deletions* in both copies of a gene that codes for a protein called *myostatin*. The deletion resulted in an incomplete protein that could not do its job. In normal cattle (and people), myostatin acts to counter the action of other proteins that drive the proliferation of myoblasts, the cells that make new muscle. Its job is to keep muscle growth in check. The mutation in the cattle causes a loss of function; in affected cattle the muscles keep growing until some other feedback mechanisms kick in to constrain growth.

Since 1997, there has been much work done exploring the so-called "double muscling" phenomenon in more than a dozen breeds of cattle; in each case, scientists have confirmed that mutations in the myostatin gene are the cause. Because of the obvious commercial implications, similar research has been done on myostatin function in mice, chickens, zebrafish, and a growing number of other animals. In 2003, a French research team showed that the explosive growth of the pectoralis major muscle (the muscle you eat when you eat a chicken breast) in young chickens was driven by the interaction of two proteins, insulin like growth factor-1 (IGF-1) and myostatin. In the United States, a team in Iowa created a strain of transgenic mice lacking a functional myostatin gene and found that these "knockout" mice had twice the muscle mass of otherwise identical mice. (Scientists create knockout animals by a technique called homologous recombination, in which a dysfunctional form of the gene of interest is injected into the early embryo. The injected gene causes the replication machinery in some cells to eliminate the normal gene and instead use the injected one.) Also in 2003, a research team in England developed an antibody against mouse myostatin and showed that, when this antibody was injected into adult mice, the animals rapidly grew bigger, stronger muscles. Researchers really do have methods for testing mouse muscle strength. (Please do not imagine miniature barbells; the scientists use strain gauges.) In essence, the antibodies disabled the circulating myostatin, thus eliminating the check on muscle growth and allowing muscles to grow bigger.

Evolution is ultraconservative; it is not a bit surprising that both the structure and function of the myostatin protein (which is also known as growth differentiation factor 8 or GDF-8) are remarkably similar in species ranging from fish to humans. For this reason, the antibody experiments in mice that resulted in increased strength have created significant interest in the possibility of developing drugs to disable myostatin. The hope is that such an intervention would effectively treat the muscle wasting that is so common in many human cancers, AIDS, and the different forms of muscular dystrophy. Antibodies or medicines that inactivate myostatin might slow or even stop muscle-wasting diseases by permitting myoblasts to proliferate far faster than normal, offsetting the loss with new muscle. There is also great interest in the possibility that such a drug would combat the natural decline in muscle strength that is part of the aging process. If such an effect could be demonstrated and the drug had an excellent safety profile, it might attract millions of users. Baby boomers have no plans to age gracefully.

The first clinical trial to study the effects of a myostatin inhibitor was launched by the Wyeth Pharmaceutical Company in 2005. It is studying the safety profile and proper dosage of an experimental compound called MYO-029 in adults afflicted with any one of three forms of muscular dystrophy. MYO-029 is an antibody (which means it is delivered by injection, not by capsule) that scientists at Wyeth designed to bind to myostatin, thus limiting its effects. It is, I think, unlikely that MYO-029 will be helpful to patients with muscular dystrophy. Their muscles are so damaged from their underlying disease that myostatin levels may play a much less important role in their bodies than in those of unaffected people. This will almost certainly not dissuade the pharmaceutical industry (nor should it). There will be other trials involving other diseases. The potential value to patients with some muscle-wasting diseases from the successful development of a drug like MYO-029 is immense.

It is also, I believe, indisputable that the potential revenues from selling an anti-myostatin drug which provides a safe means to enhance muscle mass in the healthy young and sustain it in the healthy elderly is many fold larger. In the United States, about 100,000 people are affected with one of nine different forms of muscular dystrophy, and perhaps another 400,000 suffer from severe muscle wasting because of AIDS and cancers. These numbers pale before the millions of people who spend billions of dollars trying to develop or maintain healthy muscles. At least one company de-voted to bodybuilding already claims that "Genetically gifted bodybuilders have very low levels of this protein (myostatin) in their bodies, which is be-lieved to be the cause of their muscle building gifts." Claims like this will likely cause a stampede of interest in off-label (illicit) use of anti-myostatin drugs if they come on the market. The fact that they will probably have to be injected will slow their use among teenagers, but, as the unfolding story about use of designer steroids among professional baseball players suggests, not among men and women who earn their living in competitive sports.

My guess is that use of anti-myostatin drugs will surface as an issue in professional sports within the next few years. It is even possible that a black market could emerge for anti-myostatin compounds long before a drug approved by the FDA is on the market. There is a growing scientific literature on myostatin, and thousands of biochemists know how to de-velop antibodies. Note that I have deliberately avoided using verbs such as "misuse" or "abuse." This is because I am not sure where I stand on the whole issue of efforts to "enhance" performance. Elite athletes devote most of their lives to trying to do just that. They follow special diets, consume expensive nutritional supplements, retain world-class fitness trainers, con-sult sports psychologists, take drugs to control pain so that they can com-pete in key events even when they are injured, and have access to the top orthopedic surgeons. Where is the line between spending more money for these purposes and using a drug to help build muscle mass?

Imagine for a moment that the little boy in Germany with the "double muscling" phenotype (the first human clearly shown to be so affected), a child whose mother was a professional athlete and whose grandfather could carry 300-pound curbstones, is encouraged from a young age to pursue com-petitive sports. Perhaps even now he is being trained in an age-appropriate way to maximize his unique potential. Suppose that as he grows into adolescence, it becomes clear that his mutations are, indeed, far more helpful than harmful. In his childhood, he will probably just be a local phenomenon—the fastest, strongest kid in the school, the star of the soccer team. But, a decade hence when he is a teenager, his life could change dramatically. Perhaps he will break Germany's or the world's record for the discus throw or the shot put. It is possible that he will not merely break the record, but shatter it, extending the record to distances not even contemplated by internationally ranked athletes now in their prime. This will earn him unprecedented international fame and immense wealth (from product endorsements). At 21 he could have the fame that Michael Jordan took 15 years to reach. Of course, with it could come unprecedented emotional pressures.

The article in the *New England Journal of Medicine* that described this child was circumspect about his identity. However, a number of physicians, nurses, relatives, and close family friends almost certainly know of his peculiar genetic condition. If his athletic prowess is remarkable from an early age, it is inevitable that journalists will become fascinated with his performance and the reasons for it. As he matures and seeks friendships with his teammates, will his parents (as in the recent Pixar film, *The Incredibles*) urge him to compete well below his ability so he can fit in with the crowd? Once the press come to know him, it is virtually certain that glowing descriptions of his feats in high school and college will be accompanied by references to his genetic condition. Will that make him a hero or a freak?

Will record-breaking performances by this gifted young athlete be marked with asterisks because of his genetic condition? Will he be a sort of genetic Roger Maris (about whom baseball officials noted that he broke Babe Ruth's home run record in a season that was eight games longer)? Or is the world of professional sports so enamored of athletes endowed with unusual physical capacities that he will just be another great first-round draft pick? After all, being "double muscled" is not really much more unusual than being 7 feet and 6 inches tall. Both phenotypes happen by chance to fewer than 1 in 1,000,000 people.

Being German, he is not likely to choose baseball as his sport. But what if it was found that he could throw a baseball 110 miles per hour with great precision? What if he was an American kid who ran off a string of 40 consecutive wins in high school, putting up an ERA of .05 and pitching 34 no-hitters? What if, at age 18, after signing a 5-year contract with George Steinbrenner (owner of the New York Yankees) for 250 million dollars, he proceeded to so completely dominate batters that he hardly ever lost a game? Would the fans at Fenway Park file a class action suit to forbid him from playing baseball? On what grounds? For how many years would people flock to watch him pitch before attendance faded because the outcome was always foreseeable? Is not the essence of sports a struggle in which the outcome is uncertain? Part of the power of Malamud's great baseball novel, *The Natural*, is that the protagonist was flawed.

One boy with mutations in his myostatin genes is not likely to alter professional sports all that much. Indeed, depending on his temperament, he might as he matures derive little pleasure from an uninterrupted string of victories against insignificant competitors. Perhaps he will not even feel passionate about sports. Maybe he will prefer to live in a small town and limit his show of prodigious strength to arm wrestling with his pals at a local pub. Maybe. It is far more likely that his family, his childhood coaches, his friends, and, eventually, the media will push him to pursue greatness in the peculiar world of elite sports. Whatever life trajectory the German boy with the oversize muscles takes, over the next decade our understanding of genes is going to revolutionize the approach to training elite athletes.

THE GENETICS OF ELITE ATHLETES

There is no question that genes greatly influence athletic ability. We have known it for a long time. Simply put, gender is determined by genes, and there is a significant gender gap in athletics. Although men are not "better" than women, they are on average bigger, stronger, and faster. Despite the impressive gains made by women in marathon running, the world's best women are still far short of the men's record. Across virtually all track events, the women's world records are at least 10% short of the comparable times and distances achieved by men. In 1992, Brian Whipp and Susan Ward of UCLA studied the records of five Olympic running events from the 1920s to 1990. Curves drawn from 1920 to about 1990 suggested that women would soon equal men in these events. However, since then the gap has not narrowed; it has widened. Many think that much of the narrowing of the gender gap that occurred in the 1980s, an era dominated by East German women, was powered by the illicit use of testosterone-like steroids. It also may be that during the decades in which women were catching up to men, they were steadily benefiting from coaching and training that they had been denied in the past. At least for some events, an analysis of the men's records also suggests that they too may have reached the limits of what they can achieve without the benefits of enhancing drugs or the luck of being born with uncommon gene variants.

One of the most fascinating aspects of elite sports is that certain events seem to be dominated by particular ethnic groups. Are there gene variations that are far more prevalent in one group of humans than in any others that strongly favor superior performance in particular sports? Yes. In 1968 Kip Keino, a Kenyan, captured the gold medal in the 1,500 meter race at the Olympic Games in Mexico City. For the ensuing four decades, Kenyans have achieved unprecedented world domination in running events ranging from middle distance to the marathon. In the last 14 years, Kenyan men have won 13 of the Boston marathons. They hold most of the top race times from 800 meters and up. There are 6 billion people on the planet, of which only about 30 million are Kenyans. Most of the great Kenyan runners are from a group of tribes called the Kalenjin, who number less than 3 million. Thus, a group from an impoverished part of the world that accounts for about 1 in 2,000 people on the planet has garnered most of the long distance running records for nearly two generations! Some have called this the greatest concentration of achievement in sports in history. What explains this astounding success?

The Kalenjin train at relatively high altitudes. This certainly maximizes the oxygen-carrying capacity of their red blood cells (high altitude induces red cell production), which is critical to eliciting maximum running performance. But many great distance runners throughout the world train at high altitudes. The Kalenjin eat a simple, maize-based diet. Is that the secret? Running is central to their culture; children often run miles to school. Does rigorous training from an early age provide the small edge that captures a world record? None of these observations explains the astounding success of the Kalenjin people. Well-trained Scandinavian runners equal Kalenjins in oxygen-carrying capacity. Danes have been shown to outperform them in the ability to maintain maximum heart rate on treadmill tests. The Kalenjin diet is rather poor in protein. Kenyan children are not on average more physically active than are northern Europeans.

Researchers have recently uncovered two facts about Kalenjin runners that do seem to explain part of their edge. As muscles tire and reach the limits of oxygen fueling, they produce lactic acid. Bengt Saltin, a Swedish physiologist, has shown that Kalenjins have about a 10% advantage over Europeans in squeezing energy from the same oxygen supply. This is in part because, on average, the calf muscles of Kalenjins are smaller and weigh significantly less than those of Europeans. They have powerful thighs with extraordinarily thin lower legs that require less energy to move. The second reason is less obvious than the shape of their legs. Many of them have an enzyme variant that causes low lactate production and rapid lactate turnover. This increases the efficiency with which their muscles work. Most sports physiologists who have studied Kalenjin runners think that they have a special advantage because of gene variants that affect the shape of muscle mass in their legs and others that influence the efficiency of cellular metabolism. In one study, Kenyans were shown to need 8% less energy to run a kilometer than Europeans. In the world of elite racing 8% is a huge advantage.

West Africans or persons of West African descent (such as the great Carl Lewis) also dominate the other half of racing—the sprints. Is this achievement also influenced by genes? Studies have shown that West Africans on average have denser bones, less fat, narrower hips, and longer legs than do whites. They have a markedly different habitus from, and tend to weigh 60 pounds more than, the great Kenyan marathoners. But the key difference between West Africans and everyone else is in muscle physiology. Research suggests that, in addition to the incredible discipline and dedication of the great West African runners, they also have a helpful gene variant that supports explosive sprinting. Scientists categorize muscle fibers by how quickly they can contract. Type I contract slowly; type II quickly. Type II muscle tissue is further subdivided into fast and superfast contraction fibers. Great marathoners have mostly type I; great sprinters have mostly type II. In a study comparing Canadian students to West African students, scientists found that the Africans had significantly more fast twitch (type II) fibers.

It would be foolish and wrong to suggest that one group of people is in some way genetically superior to other groups. I am not suggesting that. I am suggesting that there is growing evidence that at the extreme margin of human sporting endeavors, there are persons who, in addition to possessing the immense dedication it takes to triumph on the world stage, may have been born with gene variants that by chance support their goals. As with all gene variants, these (and those yet to be discovered) will be found among all human groups. However, the frequency with which the variant is present might vary widely.

The first use of genetic information in elite sports may well be as a tool to screen top young athletes to determine whether or not they have a genetic condition that could endanger their health or even their lives. A case unfolding as I write this chapter is illustrative. Eddy Curry, a young center signed by the Chicago Bulls, missed the last month of the 2004–2005 season after he was diagnosed with an abnormal heart rhythm. One of the team doctors has advised that he have a DNA test to determine whether he has a genetic condition called hypertrophic cardiomyopathy, the disorder that killed Celtics' star, Reggie Lewis, 12 years ago. Curry has refused to take the test, and the players' union backs him. But it seems likely that if he does not take the test, the Bulls will release him when his contract expires at the end of 2006, and that his dreams of a $70 million mega-contract are over.

This situation anticipates a rapidly approaching future that will raise many tough questions. The collective bargaining agreement that governs the relationship between players and teams gives the team the right to compel the player to submit to all "requested examinations and tests." Should the Bulls have the right to compel Curry to be tested? What if they do not test him, allow him to play, and he does drop dead during a game? Is the team liable? If he does have the genetic disorder, should Curry be allowed to assume the risk? Should teams have the right to screen all their players for disorders like this one?

The use of DNA tests to identify persons for unknown health risks will someday be routine, justified by the legitimate desire to avoid a preventable death. This will be a step toward a future in which coaches use genetic testing as part of the process to assess who may have a genetic profile suggesting that a particular individual is more likely to reach a higher level of performance than other apparently equally talented athletes in a particular sport or event. In the world of elite athletes, tiny advantages differentiate those who will merely perform superbly from those who will reach Olympian levels of performance. As the following research suggests, early versions of such screening could be rationally undertaken in the near future.

The renin-angiotensin system (RAS), a series of enzymes including one called angiotensin-converting enzyme or ACE, plays a critical role in the circulatory system. Among humans there is a relatively common variant in the underlying gene called the I variant (it is an insertion of 287 extra DNA bases). In a number of studies comparing certain types of elite athletes to normal individuals, the I allele has been found with a much higher frequency than would be expected by chance alone. The groups include long distance runners, rowers, and mountain climbers who assail high peaks without supplemental oxygen. In one study of Spanish athletes, the I allele occurred at such a high frequency among the athletes as compared to the controls that the P value (the possibility that this occurred by chance) was .0009. Interestingly, some studies show that the distributions of the I allele and its counterpart, the D (for deletion) allele, vary dramatically by particular sport. Generally, athletes focused on burst activities such as any kind of sprint were more likely to have the D allele, whereas endurance athletes were far more likely to have the I allele. Current research suggests that the major impact of these variants is realized in the metabolic efficiency of individual muscles rather than in a systemic effect on respiration. Although the data are mixed, the available body of evidence could motivate savvy Californian or Australian swim coaches to test all their promising young teens for the I/D polymorphism and use the results as part of their calculation in matching them to certain events.

Such speculations are not limited to a single gene. Haptoglobin, a large protein that was first identified in human serum more than 60 years ago, has a number of functions. The most important is as an antioxidant. It readily binds hemoglobin, which if released into the bloodstream due to breakdown of red blood cells, is toxic to other tissues. Among humans, there are two major forms of the proteins that manifest as three major phenotypes (1-1, 2-1, and 2-2). Each of these has a somewhat different level of efficiency in carrying out its biochemical tasks, and researchers have found that types 1-1 and 2-2 are associated with different risks for certain kinds of vascular disease.

Because of the important (albeit indirect) role that haptoglobin plays in the efficiency with which the body's system for delivering oxygen to tissues operates, scientists have begun to study whether the various forms of the protein are associated with athletic performance. For example, a recent study of elite athletes in Korea found that those who were type 1-1 had a higher oxygen-carrying capacity. Although it did not reach statistical significance, an excess of 1-1 was also found among marathon swimmers.

The neurophysiological subsystems (such as breathing, heartbeat, and sleeping) required to maintain human life operate within the autonomic nervous system (and endocrine glands), without any help from cognitive functions. These subsystems are regulated by two major neurotransmitters released from nerves and taken up by several types of receptors on the surface of the cells that they target. The receptors themselves fall broadly into two classes, called alpha and beta. The beta-receptors are mainly active in controlling heart function; the alpha-receptors have a much broader set of duties.

The noradrenergic system has long been associated with behaviors that are, at least from an evolutionary perspective, key to survival. These include rapid arousal, vigilance, hostility, and formation and maintenance of certain kinds of memory. Because this system is so closely related to fundamental survival strategies, researchers in Canada investigated whether there were variants within the genes coding for the key receptors that were associated with superior performance in some aspect of athletics. They asked whether two relatively common variants, one in the alpha-2A-adrenoceptor gene (ADRA2A) and the other in the beta-2-adrenoceptor gene (ADRB2), were more or less common among elite endurance athletes as compared to sedentary controls. They found modestly impressive evidence that variation in the ADRA2A gene was associated with the ability to sustain endurance training.

Perhaps the most intriguing research to date concerning genetic predisposition to success in sports is a study done in Australia of a gene, ACTN3, which is one of two genes coding for proteins called skeletal muscle alpha-actinins. The ACTN2 protein is found in all muscle fibers, whereas the ACTN3 protein is found only in the so-called "fast twitch" fibers. In 1999, the Australian scientists reported that about 20% of persons of European descent have two copies of a mutation that prevents them from having functional ACTN3 proteins. Apparently, this deficiency does not cause a muscle disease because ACTN2 protein can make up for the loss. It turns out that the prevalence of the mutation varies from a high of 25% of Asians to a low of less than 1% of African Bantu people. This and other facts led the scientists to postulate that variants in the ACTN3 gene may confer survival advantage depending on the environment. They reasoned that the genetic variants might also be associated with muscle function. This led them to study the distribution of the genotypes among 429 elite Australian athletes and 426 unrelated, nonathletic, healthy control persons. Elite athletes were defined as those who had competed on the national level. The group included 50 who had qualified to represent Australia in the Olympics.

On the basis of their understanding of the probable role of ACTN3 protein in muscle function, the scientists hypothesized that persons who lacked functional protein would be less able to compete well in sprint and/or power events, but might be favored in endurance events. When the distribution of ACTN3 genotypes among elite athletes as a whole was compared with the distribution among the controls, there were no differences. However, when the athletes were divided into those who performed in sprint/power events and those who competed in endurance events, and each was then compared to the controls, the results were markedly different. Sprint athletes had a much lower frequency of the so-called XX phenotype (lack of functional

ACTN3 protein). In fact, not one female elite sprinter had this pattern, whereas 18% of controls did. On the other hand, there was a higher frequency of the XX genotype in endurance athletes. Overall, the study suggests that the ACTN3 protein provides an advantage for power and sprint activities. The researchers speculated that it has been evolutionarily optimized to minimize damage to muscles undergoing sudden contraction.

Is it possible that during the course of human evolution, natural selection has fostered a trade-off between speed and endurance? This notion is certainly compatible with data which show that among competitors in the decathlon, performance in events that depend on sudden power generation (shot put, long jump, 110-meter hurdles) correlate *inversely* with performance in the 1,500-meter race. It is a rare athlete, indeed, who excels in both. The scientists who did the ACTN3 study have suggested that genetic testing might be a useful predictor of athletic performance at the elite level.

What do such discoveries mean for the future of world-class amateur and professional sports? How long before college coaches will be ordering up gene tests as part of their assessment of whom to recruit for their teams? Will new companies emerge to sell gene chips that scan for hundreds of variants thought to confer certain advantages in sports? Will there be a "distance runner chip," a "swimmer chip," a "boxer chip"? If there were, how much would they really change the world of sports?

Each year, geneticists at Laval University in Canada publish their update of loci on the human gene map that have been associated with athletic performance and health-related phenotypes. In 2002, the map included 90 nuclear genes and 14 mitochondrial genes. Many of these selections are based on relatively limited data, but it would not be foolish to think of them as version 1.0. In 2005, another research group reported that in a gene expression study of young men who were intensively training for cycling competition they found 500 genes (among our 25,000) that seemed to be most relevant to fitness. One can expect many more genetic studies of fitness over the coming years.

Already, one can discern three scientific trends that will propel genetic screening of athletes: interest in how gene variants affect human physiology (discussed above), pharmacogenetics, and nutritional genetics. Pharmacogenetics—the study of how genes influence response to drugs—will in time greatly refine both the medical and the illicit use of drugs. The question of which (currently) illicit steroid is most likely to help build muscle in a particular individual may someday be answered by a simple DNA test!

Nutritional genetics—the science of choosing optimal nutritional supplements to enhance health and performance—is in its infancy, but there is every reason to expect it to mature. Perhaps someday in the not-so-distant future, top athletes will adhere to diets based on computer analysis of their genetic profile. There will be a new level of science to the training table.

It is only a matter of time before genetic screening of young, potentially elite, athletes is used as part of their appraisal by elite coaches and training camps. The argument for so doing will be especially persuasive at the elite levels of each sport, for in that rarified world, slight, but unalterable, advantages and disadvantages could mean the difference between having a collection of clippings from the local newspaper versus making an Olympic team. After all, everybody else will be doing it.

Not so very far off (my guess is less than a decade) is the possibility that some top athletes will turn to gene doping to enhance performance. By injecting extra copies of genes known, for example, to maximize muscle physiology, athletes might gain an edge that would be almost impossible to detect. It is difficult to assess the physical risk of gene doping, but it is not trivial. In just a few years after the blood-forming drug, EPO, became available on the black market, 18 young European cyclists died of heart attacks (a risk of that drug). To screen for "genetic cheating," those who regulate sporting events would have to require expensive and painful muscle biopsies. High-tech labs would have to extract the DNA from the cells and sequence portions of it, looking for subtle signs that a slightly altered version of the natural gene was incorporated in the cell's genome. Eventually, detection technology would catch up with cheating technology. Until then, we may compete in an era in which blood doping will be seen as archaic compared with the wizardry of genetic cheating.

The Laetoli footprints are the oldest known footprints of our ancestors—about 3,000,000 years old. (Photo by Peter Jones; reprinted from Johanson D.C. and Edey M.A. 1981.)

Our Ancestors

One of my favorite works in the Museum of Fine Arts in Boston was painted by Paul Gauguin in Tahiti near the end of his erratic life. This large, rich-hued work portrays a dozen people captured in quotidian moments amidst the lush vegetation of the South Pacific. On the lower right three women admire an infant; far to the left an old woman seems to be slowly walking into darkness. Hidden in the shadows behind her stands a life-sized Buddha. On the upper left of the canvas Gauguin wrote three questions: "D'ou venons nous? Qui sommes nous? Ou allons nous?" Where do we come from? Who are we? Where are we going? I first saw the painting more than 40 years ago, and I have visited it many times since. When I was a college student, it seemed alluring and exotic. Now, from the perspective of more years than Gauguin had when he finished it, I find the scene solemn and the figures wistful and acquiescent. They seem to tell me that it is our fate to contemplate unanswerable questions. Perhaps so. But, thanks to advances in anthropology and genetics, at least one of these three deep questions is answerable. As we start the 21st century, we have a pretty good idea about where we came from.

The answer contradicts the origin myths of all pre-scientific societies. It also rejects the beliefs recently held by modern societies. Until about 1800, most of the civilized world—Europe, the Americas, the Middle East, and other Judeo-Christian and Islamic domains—*knew* that all humans were descendants of Noah's three sons and their wives, the only human survivors of the great flood. But as early European explorers brought home exotic animals (especially primates), and as comparative anatomists studied them, it became steadily more difficult to think of man as disconnected from the rest of the mammals. Well before the appearance of Darwin's great book (1859), scientists had grasped the concept of common descent from ancient species. As early as 1809, Jean Baptiste Lamarck included mankind in his general theory of evolution. In 1844, Robert Chambers anonymously published *Vestiges of the Natural History of Creation*, in which he argued at length that mankind occupied the highest rung of an *evolutionary* ladder. Darwin cautiously avoided questions about human evolution in *The Origin of Species*, but even the most obtuse reader must have sensed the implications that the theory of evolution by natural selection held for humanity. Although millions of Americans still cling to the belief that God placed Adam and Eve in Eden in 4004 B.C. (a date that 19th-century Biblical scholars calculated by analyzing the genealogies in Genesis), by the mid-19th century discoveries in geology and paleontology had provided overwhelming evidence that both the earth and its inhabitants were of much greater antiquity. Discoveries of fossilized human remains were crucial to the recalibration of planetary history and the construction of our family tree.

THE BONES OF OUR ANCESTORS

On four occasions between 1828 and 1899, workmen or amateur naturalists uncovered the fossilized bones of strange humans with heavy brow ridges, thick femurs, and large craniums. In 1856, workmen discovered a skull cap and other bones near the Neander River close to the city of Dusseldorf in Germany. Thinking they had found a murder victim, they brought the material to a local science teacher, J.C. Fuhlrott, who startled everyone by saying they were not the remains of a modern man. They were next taken to a leading German anatomist named Schaaffhausen, who also concluded that they belonged to an extinct race of man. Over the next few years, most of Europe's leading anatomists rendered their views. One opined that the skull was the remains of a Cossack soldier who had fought against Napoleon. Another suggested that it was a modern man who had been born with several congenital abnormalities. The press preferred that the remains be those of a mysterious, ancient race. It dubbed the people they typified the Neanderthal.

The subsequent discoveries of similar bones in Western Europe (France, Germany, and Croatia) suggested that a human quite like, yet distinctly different from, modern humans had preceded us in Europe. The idea that humans had originated in Europe resonated well in the 19th-century European countries ruling far-flung colonies that they had already claimed were populated by inferior peoples. There were, however, scientists (Darwin among them) who speculated that early man must have evolved in a warmer climate, and that Africa or Asia was a more likely ancestral home than was Europe.

Much of 19th-century European geology was devoted to understanding the impact of climate change on the fluctuating diversity of animal species and on the rise of modern humans on that continent. The focus was on events of the Pleistocene, a period of about 700,000 years during which temperature fluctuations drove four periods of glacial advance, each followed by a period during which the ice receded (we are in the fourth one now). During this time, the climatic effects were profound. Sea level rose and fell as much as 300 feet. At the height of the glacial advances toward the equator, tropical rain forests became grasslands and grasslands became desert. Most animal species (especially the larger ones that have huge energy needs) either made dramatic adjustments or died.

It was during or shortly after the last of these glacial retreats that the first great cave paintings were created by humans in the Dordogne in southern France and in nearby Spain and Italy. In the same caves, scientists found innumerable fossil bones. Most were the remains of the beasts the cave dwellers had hunted, but some were the bones of the hunters themselves—humans who were tall and slender, a race called Cro-Magnon (named for a village in France near where they were uncovered).

As dating techniques improved, scientists realized that, at least in Europe, Neanderthal man coexisted for thousands of years with Cro-Magnon man. This posed a difficult problem that dominated anthropology for a century: How were the two human species related? Until about 1890, the most popular solution was to posit a straight lineage. Most paleontologists thought that Neanderthal preceded and gave way to Cro-Magnon, who in turn gave rise to the modern humans. However, more recent evidence suggesting that the Neanderthal died out about 28,000 years ago made most scientists conclude that these low-browed people were a mere side branch of humanity, an evolutionary dead end. Even if we shared a distant ancestor, our more immediate ancestors had either extirpated or outlasted them. That they lost the battle of the survival of the fittest should not lead one to imagine Neanderthal as unintelligent. An ample scientific literature records and analyzes their tool-making skills.

Given the strong evidence that the two human species lived in Europe contemporaneously, could Neanderthal genes have flowed into the lineages that gave rise to modern humans? As early as 1916, Henry Fairfield

Osborn, a prominent paleontologist who was president of the American Museum of Natural History, argued such gene flow, if any, was "accidental" and inconsequential. He asserted that there is "no recent type which can be considered even as a modified direct descendant of the Neanderthals." That remains more or less the dominant view today. At a meeting of Neanderthal experts held in New York in 2005, the consensus (based in part on observations of matings between closely related mammals) was that there was probably some gene flow, but that it today accounts for a minuscule part of our genome. The few mitochondrial DNA studies of Neanderthal bones support this; their DNA is quite distinct from that of modern *Homo sapiens.*

The question "Where do we come from?" had to be reconsidered after 1891. It was then that Eugene Dubois, a young Dutch surgeon who was passionately interested in fossil hunting, joined the Dutch army for the sole purpose of obtaining a post in Java, where he theorized that fossil hunting would yield rich rewards. Heavily influenced by the views of Ernst Haeckel, the great Swiss paleontologist who had postulated an evolutionary connection between humanity and the great Asian apes, Dubois was convinced that *Homo sapiens* had evolved in a warm, wet climate. Within months of his arrival, the young surgeon (probably the luckiest fossil hunter who ever lived), investigating a riverbank near a place called Trinil, found two hominid molars, a skull cap, and a left thigh bone. In 1894, after two years of painstaking study, Dubois presented them to the world as evidence of an early human that flourished hundreds of thousands of years *before* Neanderthal. The skull cap permitted the inference that the remains were those of a hominid with a brain case about half the size of that of modern man, and the leg bone strongly supported an inference of upright posture. A comparatively big brain and bipedality are the two key criteria to support a claim to being an ancestor of humans. Dubois named the early man *Pithecanthropus erectus* (literally, erect ape man). Years later, he changed his mind, arguing that his find was an ancient great ape, perhaps an ancestor of the orangutan. But, by then, other scientists were convinced that *Pithecanthropus* (today called *Homo ergaster*) was part of the human ancestral tree, however distant.

Studies of *Pithecanthropus* weakened the notion that *Homo sapiens* arose in Europe. Yet, *Pithecanthropus* was so much more ancient and so physically different from *Homo sapiens* that many scientists viewed him as an ancient side branch of hominid evolution (rather than a direct ancestor). This was reinforced in 1927 when a Canadian biologist named Davidson Black discovered a fossil hominid tooth near Peking that he audaciously claimed belonged to a distinct hominid genus and species that he named *Sinanthropus pekinensis*. As other discoveries of *Sinanthropus* emerged, evidence mounted that it was the same as *Pithecanthropus*. Although a few scientists postulated that *Pithecanthropus* represented a line that had arisen in Europe, the much older age of the Asian countered their argument. After World War II, scientists renamed these Asian fossils *Homo erectus*.

Perhaps the most dramatic discovery of the first half of the 20th century occurred in 1924 when Raymond Dart, an Australian anatomist who was a professor at the University of Witwatersrand in Johannesburg, reported on a fossil hominid that a miner had discovered in the Taung limestone quarry in South Africa. Dart himself chipped away the rock, exposing a nearly complete skull of a young child. Based on the position of the hole in the base of the skull through which the spinal column passes, Dart asserted that this early hominid walked upright. He named it *Australopithecus afarensis* (southern ape man). It should be noted that the word has nothing to do with Australia. With an estimated age of 2.5 million years, the Tuang fossil pushed hominid evolution much, much deeper into the past.

Unlike the course taken by most scientists who often convened a team of experts to study new finds before publishing them, Dart (who was not known among anthropologists) was quick to publish. His paper, which appeared in *Nature*, generated a storm of controversy. Most anthropologists thought that the low cranium, prognathic jaw, and lack of chin were those of an ancient ape, not a hominid. Not until the subsequent discoveries by Louis Leakey (one of the most colorful figures in the history of fossil hunting) of other extremely ancient hominids in the Olduvai Gorge in Africa beginning in the 1930s and continuing for four decades did the majority accept Dart's point of view. More than any other man, Dart was responsible for the late 20th-century consensus that hominids, the ancestors of man, arose in Africa. Fittingly, this blow to 19th-century European conceit came as the colonial era was beginning its decline.

Arguably, Leakey's greatest moment came in 1962–1964 when he and his colleagues, John Napier and Philip Tobias, found and described the fossil remains (three skulls) of a new creature with features much more human-like than those of the Tuang fossil that they named *Homo habilis* ("handy man"). The scientists dated the remains as being 1.75 million years old, which suggested that their discovery had tripled the age of the lineage that they could confidently connect to modern humans! But was *Homo habilis* truly an early human ancestor, or was he a hominid representing a side branch of our tree that had died out? The controversy about the discovery centered on a single question: What size brain differentiates early humans from non-human apes? The *Homo habilis* skulls have a mean brain capacity of 642 cc. This is less than half the size of a modern human adult brain, but it is 200 cc larger than the brain volume of the australopithecine skulls. Of course, all such dividing lines are arbitrary. But the habilis brain volume is so much larger than that of the australopithecine that many thought they were not merely different species, but that they were even of a different genus.

One of the next great moments in our quest to decipher human origins took place in Hadar, Ethiopia on November 30, 1974. A young American scientist, Donald Johanson, and his colleague, Thomas Gray, were exploring a gully when they discovered an unusually intact set of remains of a small hominid. A combination of excellent stratigraphic data and sophisticated chemical analysis later allowed them to date the fossil age at 3.5 million years. The two most important aspects of Lucy—her great age and the completeness of her skeleton—make her unusually important in the analysis of human origins. Lucy was 3 1/2 feet high and weighed about 60 pounds. She walked upright, she was powerfully muscled, she had human-like hands, and her brain was about the size of a chimpanzee brain. Put crudely, Lucy looked more human from the neck down. Today, a majority of experts believe that Lucy (and related skeletal remains found in subsequent years at Hadar) is in the ancestral line from which *Homo* evolved, but the question is not yet fully resolved.

A surprising development occurred in 2001 when workmen excavating medieval ruins in Dmanisi, a town in the republic of Georgia (part of the former Soviet Union), found several largely intact early hominid skulls. This capped a decade of work during which Georgian scientists had found several partial skulls along with many simple stone tools that they determined were about 1.8 million years old. The Dmanisi skulls are small, but the hominids they represent walked upright. The Georgian finds suggest that an early member of the genus *Homo* (it is uncertain whether it is more closely related to the older *H. habilis* or the more recent and larger *H. erectus*) emigrated from Africa as much as a million years earlier than had been previously thought. Among several possible scenarios raised by this discovery is that the ancestor of man migrated out of Africa long before *Homo sapiens* existed, and that one of its descendants later returned to Africa to become the immediate progenitor of modern humans.

We can confidently expect that the search for human origins will yield new findings and spark new controversies for decades. Just a few years ago the discovery of the remains of *Homo erectus* along the Solo River in Java that were only 25,000 to 50,000 years old forced the dramatic conclusion that for a time three different human species (*erectus, neanderthalensis,* and *sapiens*) coexisted on earth. In 2004, Mike Morwood and a team of scientists working on the Island of Flores in Indonesia made the astounding discovery of the remains of a tiny human that they believe can claim membership in our genus. Called *Homo floresiensis*, our unusual cousin stood well under 4 feet tall and had a brain about the size of a chimp's (380 cc). Its discoverers believe that *H. floresiensis* is a direct descendant of *H. erectus* who arrived on the island about 800,000 years ago.

The claims being made about *H. floresiensis* are believable. Evolutionary biology has many examples of species undergoing rapid change in size once they become geographically isolated. The most important, of course, is that there are in Africa today groups of human pygmies, for example, the Bambuti. Another of my favorite examples is the miniature horse, a beautifully proportioned animal that stands only 30 inches high. Although records show that miniature horses emerged in Europe and were bred to be the pets of royalty (such as the 17th-century Hapsburgs), the separate emergence of the Falabella horse in Argentina is better documented. Members of the Falabella family first found the miniatures (descendants of horses brought by the Spanish invaders) among herds owned by the Mapuche people in southern Argentina in the mid-19th century. About 1940, Julio Falabella converted a family hobby into a major breeding program. Today these delightful animals have spread across the globe. There is even a breeding program operated by the sisters of the Monastery of St. Clare in Brenham, Texas!

In 2008, some anthropologists are disputing that *H. floresiensis* represents an extinct human line, asserting instead that it represents the more recent remains of a microcephalic member of a tribe of modern pygmies that live nearby. Certainly, its very small cranium (which is no bigger than a chimpanzee's) does make the claim that the species is part of our genus difficult to accept. However, the team of scientists that has conducted the most detailed studies of the crania of a variety of hominids did conclude that the best explanation of the data was that *H. floresiensis* was in fact a member (albeit, very small) of our genus. Early members of our species seem to have triumphed over *Homo neanderthalis* in Europe and *Homo erectus* in Asia. If *Homo floresiensis* is an extinct human that inhabited Flores until a mere 13,000 years ago, it too could have perished at the hands of our immediate ancestors. It will take years to resolve whether the remains of *H. floresiensis* are from a small, modern human or an extinct cousin.

How far back will we be able to trace our ancestry? It is pretty well established that the hominid line which gave rise to modern humans diverged from the line that also gave rise to the modern chimpanzees about 7 million years ago. Until recently, firm evidence of "obvious" hominids did not extend much beyond four million years. In 2002, a team of French paleontologists found a skull and other bones (*Sahelanthropus tchadensis*) in Chad nearly 7 million years old. The skull, commonly called Toumai (which means "hope of life" in the local language), has engendered great controversy as to whether it is a true hominid or a great ape. Recently, the use of computer techniques to recreate three-dimensional images of the skull has pulled some of the skeptics toward the hominid view.

When I muse about our most ancient ancestors, I think not of skulls, but of feet. Embedded in a field of volcanic ash that fell to earth more than 3,000,000 years ago in a Tanzanian valley called Laetoli, there are a pair of distinctly human footprints. They run side by side for nearly 80 feet before they disappear. Although there is no way to determine with certainty the gender of the two hominids who made them, I find it pleasing to think of the pair as a man and a woman walking hand in hand, perhaps to a waterhole or to a stand of sheltering trees. Perhaps one carried a young child. These impressions may be the earliest evidence of a family of our hominid ancestors we will ever have.

RECENT MIGRATIONS

Efforts to trace the human diaspora—the path taken over the last 50,000 years or so by groups of modern humans as they spread across the planet—must turn to more subtle evidence than the shapes and age of early hominid fossils that have been used to reconstruct our hominid lineage. During the mid-20th century, population geneticists used new techniques in blood group typing and enzyme analysis to develop biochemical profiles of populations. They called these "serologic races." Few markers were found only in a single race or ethnic group, but by typing different populations to ascertain the distribution of 10–20 markers, the scientists could develop a modest likelihood estimate as to the population group from which any randomly selected blood sample derived. By comparing the magnitude of the differences in the distribution of these markers among populations, we can map their biological and temporal divergence. With advances in DNA science, we can make reasonable estimates of the race of the individual from which a sample was obtained and of the relative closeness of the groups, but they are still only estimates.

Mitochondria are small circles of DNA that exist in hundreds of copies in the cytoplasm that surrounds the cell's nucleus. Mitochondrial genes code for proteins that control energy transport in the cells. They pass only from mother to child (eggs carry lots of mitochondrial DNA; sperm do not), so mitochondrial variations define maternal lineages. Over the last 20 years, molecular biologists have shown that differences in the mitochondrial DNA sequences in different populations can be used to determine the rough timing and order in which the ancestors of current peoples diverged. Such studies confirm that modern humans left Africa about 55,000 years ago and quickly moved through Asia, colonizing Australia about 45,000 years ago, and that some of our ancestors entered Europe from Asia also about 45,000 years ago. Somewhat later (between 40,000 and 15,000 years ago) other ancestors crossed the land bridge to North America. No archaeological site in the Americas has been shown definitively to be older than 14,000 years, but DNA studies are compatible with an earlier entry.

Population geneticists have used variations in the DNA sequences of Y chromosomes, which can only be transmitted from male to male, to reconstruct the recent history of human populations. Of the many new insights about gene flow that derive from Y chromosome analysis, none is more fascinating than the genetic legacy of Genghis Kahn. Born in 1167 to a powerful family in central Mongolia, at age 9 the boy (then called Temunjin), who would become one of history's most feared warriors, lost his father when rival tribe leaders murdered him. Legend has it that he and his mother barely survived the first winter after his father's death. As a teenager, Temunjin showed great physical and emotional strength, and before the age of 20 he was forging tribal alliances, one through marriage. When his wife was kidnapped by a rival tribe, Temunjin organized a successful attack upon it, saving her and forcing its leaders to swear loyalty to him. By the age of 25, he had united scores of rival Mongol tribes under his leadership. At a tribal council meeting in 1206, he was chosen as the supreme leader and given the name Genghis Kahn ("universal lord").

Historians universally agree that Genghis Khan was a military genius. He organized a large army composed entirely of cavalry and invented what the Germans called "blitzkrieg" or "lightning" warfare. Conquered peoples either joined him or were systematically executed. Between 1206 and 1215, Khan subjugated China, and by 1218 he ruled the Korean peninsula. In 1219, in retaliation for attacks on Mongol trading caravans, his armies conquered vast territories, including northern India, Pakistan, and the Middle East. At the height of his rule, just before his death in 1226, Khan ruled the largest empire (stretching from Persia to Korea) ever assembled. In an uneasy alliance, his four sons by his principal wife extended the empire. In 1235, they and some of Genghis's grandsons, concerned to protect the borders of the fiefdom granted to one of them (Batu) in the Middle East, invaded Europe, conquering much of modern Bulgaria and Poland. In 1241, they were nearing Vienna when news of the death of one of the khans pulled them back to struggle over issues of succession.

Contrary to the more lurid histories, Khan, although merciless in war, was benevolent in peace. His dynasty was famous for its tolerance of all religions. He placed his sons, grandsons, and cousins as rulers in many territories, ordering them to allow conquered peoples to control most of their destiny. Not unlike the Roman Empire in its prime, the Mongol dynasty realized that collecting taxes made much more sense than mass executions. Although fractionated by sibling power struggles, the empire that Khan established survived for about 150 years after his death, in part due to the familial ties of the various regional rulers. To this day, parts of the former Soviet Union, Mongolia, and some peoples in China still revere him.

Recently, a consortium of Asian geneticists has discovered extraordinary molecular evidence of the success of the Khan dynasty. They examined the status of DNA markers at 32 different locations on the Y chromosomes of 2,123 Asian men who belonged to more than 50 different regional or ethnic groups stretching from Armenia and Georgia in the west to Pakistan in the south and Japan in the east. Most of the men had a unique DNA code, and the few patterns common to a small group of the test subjects were generally common to their local population. However, one DNA pattern (haplotype) occurred at an unusually high frequency among closely related lineages. To the surprise of the scientists, this haplotype was found in 16 separate populations scattered across a large swath of Asia.

By studying the haplotype differences among populations, the scientists deduced that the most recent common ancestor, the male who seeded these populations with this haplotype, lived about 900 years ago. Molecular analysis indicated that the haplotype, which originated in a single family in Mongolia, is now present in 8% of all males in a region stretching from northeast China to Uzbekistan. Put another way, in the space of about 1,000 years (some 40 human generations), one haplotype was spread to 16,000,000 men, about 0.5% of those on the planet. There are only two possible explanations for this. The biological one is that this particular form of the Y chromosome DNA conferred some sort of special survival advantage (such as resistance to malaria or plague), but given the paucity of genes on the Y chromosome, this is extremely unlikely. The cultural explanation is that the man who originally carried this haplotype and his descendants long enjoyed a superior advantage in access to women. Genghis Kahn and his many descendants enjoyed just such an advantage during a dynastic period that lasted nearly two centuries! For example, Marco Polo, who claimed to have met Kublai Khan, grandson of Genghis and ruler of China, reported that he had four wives who bore him 22 sons (and an unknown number of daughters), as well as innumerable concubines. Few would dispute my suggestion that many men would be delighted to learn they are direct descendants of Genghis Kahn. At this writing, at least one pub has hit upon the novel idea of offering Y chromosome analysis to its regular patrons whose ethnic history predicts about a 1 in 12 chance that they will have the warlord haplotype!

The techniques of haplotype analysis of mitochondrial DNA and Y chromosomes are also being used for the more sober purpose of helping African-Americans and South Americans of mixed ancestry to determine the region of Africa from which their ancestors were sold into slavery. Historians estimate that over a period of two centuries about 13 million Africans were placed in slave ships bound for the new world. About 2 million perished on the journey. Ships' manifests, bills of sale, and other documents suggest that 8 million slaves came from western Africa, 4 million came from west central Africa (Cameroon south to Angola), and about 1 million came from the southeast (Mozambique and Madagascar). In March of 2004, a team led by Anotonio Salas, a forensic geneticist in Spain, reported on their detailed efforts to develop a database to help establish ancestry. They studied the mitochondrial DNA sequences of nearly 500 persons of recent African ancestry, comparing them to DNA databases compiled earlier on cohorts of more than a score of African peoples. Their results indicate that interested persons could use DNA analysis to determine broadly from what region of the continent their slave ancestors came. It should be possible to greatly improve on the African databases so that one day soon an African-American might be able to find out with a high degree of likelihood to what tribe his ancestor belonged.

Remembering the fascination of African-Americans with Alex Haley's book, *Roots*, one population geneticist has created a database of DNA variations on the Y chromosome that are most commonly found in the dozen or so modern West African nations (such as Mali, Benin, and the Ivory Coast) where the slave trade flourished in the 17th and 18th centuries. For a fee, he will test the DNA of an interested party to see whether his Y haplotype is strongly associated with a particular population in Africa. This is an emotionally risky business. About 30% of the gene variants carried by modern African-Americans derive from rape by white slave owners. Men who think of themselves as of African origin have a good chance of being told that their Y chromosome can be traced to England or Scotland.

Using DNA tests to probe ethnic background can yield surprising results. As part of a popular sociology course at Penn State, students have the option of submitting a sample to DNAPrint Genomics, Inc., which claims it is able to tell them within a percentage point the fraction of their ancestry that is European or African. According to Professor Samuel Richards, many of the white students hope to find out that they have a little bit of African blood (because they think it will upset their parents). On the other hand, some light-skinned black students seek affirmation that they are mostly of African origin. One black student who thought she was 50% African was surprised to learn that she was only 42%. Yet, she and other black students were quick to agree that being more than 50% white meant little in a cultural context. They readily identify themselves as black and, just as readily, assert that society would not permit them to claim to be white.

The explosive interest in DNA testing as a genealogical tool has ignited a vitriolic debate in France. In October of 2007, an amendment to a new immigration bill was proposed to verify, through DNA testing, the ancestry of persons who were claiming the right to take up residence based on family ties. Although eleven other European countries have adopted such a measure, many in France are bitterly opposed, saying it reminds them of Nazi racism. The crucial argument against the proposal is that, in France, the legal definition of family is not based on genetic parentage, but on declaration that a child is one's own. Those opposed argue that it would institutionalize one standard for native French and a harsher one for immigrants. The fact that DNA testing would be voluntary and used only as a tool to assist individuals making claims that could not be verified by public documents has left about half the country unconvinced, with some ministers accusing the government of racism and threatening to resign.

Interest among African-Americans in exploring their African roots has mushroomed over the last few years. Almost certainly, the key driver of this interest was the four-hour PBS special, "African American Lives," created by celebrated Harvard professor Henry Louis Gates, Jr., which first aired in February of 2006. Noting that there is no Ellis Island for black people, Gates successfully recruited a number of prominent African-Americans to permit him to use all the tools he could to pursue their genealogies as far back in time as possible. Pre-eminent among them was Oprah Winfrey, whose family history commands a major portion of the airtime.

In trying to penetrate the wall of oblivion that slave traders created during the Middle Passage, Gates sought to use DNA tests to help connect his colleagues with their African roots. Not surprisingly, efforts to connect today's African-Americans with the tribes of their ancestors forged few definitive connections. Along the way, Gates realized that many of the companies that were offering to trace African roots could not really do so. There are two main reasons. The first is that many tribes have not been distinct social entities for a long enough time to permit the emergence of DNA markers on either mitochondrial DNA or the Y chromosome that can function as a distinctive genetic profile. The second is none of the ancestry testing companies have compiled a large enough African DNA database to maximize the likelihood of making an accurate match. In effect, when such efforts are scrupulously undertaken, they produce data indicating that one could be a descendant of any one of many tribes scattered across modern Africa.

Gates' interest in DNA testing dates back to at least the year 2000, when a company to which he submitted his DNA sample told him that his maternal line most likely descended from Nubian ancestry, a black people in southern Egypt. Yet, in 2005, when Gates sought confirmatory testing, a second company told him that his maternal line descended from a European (white) woman. After studying the field, Gates concluded that many ancestry testing companies were reporting out more specific information than the data justified. To paraphrase him, they were telling people what they thought they wanted to hear.

In 2007, to much fanfare, Gates teamed up with one of the largest ancestry testing companies, Houston-based Family Tree DNA, to launch a new effort on behalf of African-Americans. He has been careful to caution about the limits of DNA technology, but he is also determined to squeeze from it as much information as possible. To that end, he has hired a team of prominent historians who can use the standard tools of historical scholarship to help customers learn the most plausible source of origin among several possibilities. No doubt, his 2008 four-hour PBS show, "African-American Lives 2" will do much to educate Americans about the role of DNA in genealogy. This is the limit on what DNA analysis can offer for now. To take the search to the next level, scientists will have to compile databases from many different African ethnic groups and subject them to detailed statistical analysis.

This can be done. Scientists recently conducted similar research among 18 different native groups in northern Asia. By sequencing 71 different mitochondrial DNA lineages and studying how much they varied among the ethnic groups, they were able to suggest the source of different population radiations over the last 10,000 years, especially to rule out a northern route for the initial colonization of Asia.

Another example of using DNA markers to resolve ancestry is the study of the genetic history of Polynesians. Among anthropologists, the standard picture (called the Slow Boat Model) is that over several thousand years, persons from East Asia migrated through New Guinea and Island Melanesia before migrating to the much more remote Polynesian islands. Surprisingly, mitochondrial and Y-chromosome DNA studies of the migration path gave sharply contradictory results. The maternal studies found that 94% of the mitochondrial DNA was East Asian, the paternal study found that 66% of the Y chromosomes were Melanesian. In 2008, a large study of 377 markers distributed across all the chromosomes found that, in Polynesians, 79% of the DNA is of East Asian origin, while 21% is of Melanesian origin. This tends to support a new view of the migration history (dubbed the "express train" model) that posits that people from Taiwan migrated through Melanesia quite quickly.

North and South America were the last major land masses to be populated by humans. The relative paucity of ancient archaeological sites, the disappearance of the land bridge between Asia and North America, the decimation of most Native American groups by European diseases, and the difficulties of linguistic analysis are among the major reasons why the arrival of the first Americans has been so hard to determine. In the mid-1980s, anthropologists, population geneticists, and linguists reached a fragile consensus. They concluded that the Americas were populated in three waves by three peoples over a period lasting about 7,000 years. The Amerinds, who populated South America and the southern half of North America, arrived 11,000 years ago; the Na-Dene peoples, who populated western Canada, arrived 9,000 years ago; the Eskimo-Aleut peoples arrived just 4,000 years ago.

During the 1990s, new genetic data began to outweigh studies of dental morphology and linguistics in estimations of the arrival and dispersion of Native Americans. Studies of mitochondrial DNA and Y chromosome haplotype analysis now strongly support a single major migration into the new world. In early 2004, anthropologists at the University of Arizona reported on their study of 2,344 Y chromosomes from 18 Native American (including representatives of each of the three linguistic groups), 28 Asian, and 5 European populations. Their data are most consistent with a single migration from an Asian homeland in or near the Altai Mountains in Southwest Siberia between 10,000 and 17,000 years ago. Their findings are only modestly inconsistent with another study that posits one major emigration (like the one they posit) followed by a minor migration (by the people we call Eskimos) from the region that once constituted the land bridge as recently as 3,000 years ago. Similar research by other groups argues for pushing the timing of the major migration back to as much as 20,000 years before the present.

The musings of Bryan Sykes, a geneticist at Oxford University who is an expert on ancient DNA, suggest that some day we may be able to reach much more detailed conclusions about our ancestral lines. Realizing that surnames originated less than 1,000 years ago, Sykes posited that he could use Y chromosome analysis to prove that most people in the United Kingdom named Sykes (a word that refers to a small stream marking a boundary) shared a common ancestor. Like surnames, Y chromosomes travel with the male line. He secured DNA samples from a random selection of the more than 10,000 Sykes in the nation. DNA analysis showed that nearly half carried a unifying Y haplotype. What of the other half? Professor Sykes (who notes he is among those who carry the original haplotype) explains that there is only one explanation—infidelity among the Sykes wives. But things are not nearly as bad as they at first seem. On average, in each generation, only about 1% of Sykes wives bore children fathered by men other than their husbands.

Perhaps the most elaborate effort to use DNA to refine genealogy was undertaken by a Utah businessman, Jim Sorenson. Through a foundation he created in 2000, geneticists affiliated with the University of Utah are traveling the world collecting DNA samples. They are using Y chromosome and mitochondrial DNA analysis to compile a database of 500,000 samples from more than 100 distinct ethnic groups. This will help those who share Sorenson's Mormon faith that includes a belief in baptizing ancestors. Sorenson (who recently died) also founded a company, Relative Genetics, that for a fee of $50 will help people who take the DNA test find to which groups in the database they best connect.

D'ou venons-nous? Here is one (unpoetic) answer to Gauguin's wistful question. The genus of which modern humans are members arose in Africa about 3–4 million years ago. Almost certainly, several species of early hominids emerged, flourished, and failed before *Homo sapiens* became ascendant. The earliest species with which we might claim a direct tie is *Homo habilis*, who appeared in Africa more than 2.5 million years ago. He had a brain larger than the australopithecines, walked erect, and used primitive stone tools. Over the next million years, *Homo habilis* disappeared and *Homo erectus* became the first of our ancestors to leave Africa. This larger hominid made it all the way to China and Southeast Asia. In time, *H. erectus* also perished. Modern humans, people you would not think looked particularly unusual if you saw them in the grocery store, distant descendants of *H. habilis*, arose in Africa about 200,000 years ago. (Two sets of fossils found by Louis Leakey and his colleagues several decades ago in the Kibish Formation in southern Ethiopia—fossils that are indisputably *Homo sapiens*—have recently been dated by several new methods as originating 195,000 years ago.)

H. sapiens probably left Africa about 55,000–60,000 years ago. He (or should I say, we) then spread rapidly across the globe, possibly following a southern route across the mouth of the Red Sea and colonizing coastal India within a few thousand years. Mitochondrial DNA analysis suggests that the wave front of migration traveled no faster than four kilometers per year. Growing evidence suggests that our ancestors colonized Australia before they moved into Europe! At some point about 40,000 years ago, some of the pioneers changed direction and, moving at a similar pace, traveled northwest to Europe. Current evidence suggests that the group that moved out of Africa was small, perhaps including no more than a few hundred women. Their descendants, the men and women of southern Europe who 20,000 years ago created the graceful cave drawings in Dordogne, France, had essentially the same artistic capacity as did contemporaries of Paul Gauguin. Mastering ocean travel took a long time. The Tahitians with whom Gauguin lived are the descendants of people from Southeast Asia who reached that lush island just 2,000 years ago.

Races of Mankind. Chromolithograph by F.E. Wright, 1896. Until the mid-20th century, many anthropologists thought there was adequate evidence for dividing humanity into many races. Modern genetic evidence squarely rejects that thesis.

Race

There are about 6,500,000,000 persons on the planet. All of us are members of the species called *Homo sapiens*, yet we differ in many ways. We live on seven continents and are the citizens of (as I write this sentence) 192 or 193 countries (depending on the status accorded to Taiwan). According to the latest edition of *Ethnologue*, an encyclopedia of languages, we speak in 6,912 tongues. In fact, most linguists assert that it is impossible to come up with a firm number, largely because of vitriolic disagreements over how to distinguish between a dialect and a language. New Guinea is the most linguistically diverse landmass on the planet. It counts 820 languages! We belong to 19 major religions—of which Christians at 33%, Muslims at 20%, and Hindus at 13% are the three largest—that scholars divide into about 270 smaller groups. From an ethnographic perspective, we are part of several thousand cultures. Currently, most physical anthropologists fit us into one of three major racial groups: Caucasoid, Negroid, and Mongoloid. Some prefer a fourth, usually called Australoid, to accommodate the aboriginal peoples who arrived in Australia more than 45,000 years ago and lived in isolation until recently.

Even though we are a single species, we do not get along at all well. According to data from the United Nations, in 2002, human beings were involved in more than 20 conflicts in each of which the death toll exceeded 1,000. In about 10 of these, the death toll ranged from 10,000 to 100,000 during that year. By several scholarly estimates, war directly caused the deaths of nearly 200,000,000 people during the 20th century. If we are indeed a human family, we are seriously dysfunctional.

The notion that *Homo sapiens* is composed of several biologically different groups among which some are superior to others is nearly as old as culture itself. In ancient Egypt, the Nubians who lived at the most southerly reaches of the Nile River were considered by the lighter-skinned inhabitants of the great delta in the north to be vastly inferior, so different as to constitute a different type of human. The most ancient scriptural rationale for racial hierarchy is Noah's curse of Ham (Genesis 9:20–27). In his anger at Ham (the founder of the Canaanites) for viewing his nakedness, Noah said, "Cursed be Canaan; a slave of slaves shall he be to his brothers." To this day, one can find scriptural literalists who cite this story to justify racial inequality (just type in the key words Old Testament and race on Google). Some scriptural literalists read the story of the tower of Babel (Genesis 11:1–9) as explaining the origin of races. Observing people building the great tower, the Lord mused, "They are one people, and they have all one language, and this is only the beginning of what they will do; and nothing they propose to do will now be impossible for them." The Lord immediately scattered the people across the face of all the earth.

From the decline of the Roman Empire until the Renaissance, there was little formal thought in the western world devoted to the concept of what became known as "race." As the apostles and those who came after them spread the gospels of Christianity, they made converts without regard to ethnicity. Certainly, the crusades gave Europeans a much greater awareness of Islamic culture, but, despite the religious wars, both sides were of the same race.

The first contacts with China (as seen through the eyes of men like Marco Polo) strengthened the notion of racial diversity but provided no basis to assert a racial hierarchy, for the Orient was in many respects both economically and culturally superior to Europe. It was only when Europeans extended their trading routes to the south that they began to construct a racial hierarchy. The European voyages of discovery in the 15th and 16th centuries along the coasts of Africa and to the New World probably played a crucial role in fostering ideas of racial superiority. Indeed, this idea was probably necessary to rationalize the growth of the slave trade that flourished from about 1650 to 1800.

Scientific interest in categorizing humanity was greatly influenced by the rise of the field of taxonomy (the classification system that all high school biology students learn as species, genus, order, family, class, phylum, and kingdom). Taxonomy sprang to life from the fertile mind of a young Swedish physician and botanist named Carl Linnaeus (1707–1778). Fascinated since early childhood with the naming of plants, Linnaeus published his first compilation, *Systema Naturae*, in 1735. During his lifetime, this slim volume grew through many editions to a multivolume work that dominated the scientific view of the relationships between living things. Linnaeus was not the first to attempt a broad classification scheme for nature (Aristotle also attempted to do so), but he made at least two significant contributions. He insisted that all organisms should be described with just two words. The first would designate the genus and the second the species. He also conceived the higher orders of organization above genera. Linnaeus was also the first to place humans in nature's great web of life, listing us among the primates with the name *Homo diurnis* ("man of the day"). Today, of course, we are *Homo sapiens* (man of knowledge). Linnaeus was also among the first to attempt a racial classification of man. He divided the human species into four groups: the americanus (native American), europeans, asiaticus (Asians), and afer (African) peoples. In sharp contrast to the physiognomic focus of the 19th and 20th centuries, Linnaeus distinguished each group mainly by temperament. For example, he considered europeans to be lively and inventive, while he viewed the afer people as cunning and capricious.

Spending most of his career as a professor at the University of Uppsala, Linnaeus trained dozens of students, of whom several became naturalists on major voyages of discovery. His most famous student, Daniel Solander, accompanied Captain James Cook on his first trip around the world. As these great voyages brought western scientists in touch with non-western peoples with starkly different appearances and cultures, it was hardly surprising that the scientists would ask whether the differences between Europeans and newly discovered peoples were so profound as to justify taxonomic subdivision.

The years from about 1750 to 1900, the heyday of colonization by the European powers, saw the emergence of physical and cultural anthropology. Western scientists made scores of expeditions into relatively uncharted regions to study the humans who lived there. Much of this work was done in Africa and Asia. It is no accident that deep interest in racial typology overlapped the century (1750–1850) in which slavery was at its zenith as an economic institution. Scientific assertions that there were distinct differences between groups within *Homo sapiens* were essential to support arguments that one major group (Caucasians) was superior to others, an assertion that was crucially important to the rationalization of slavery in a modernizing world.

As anthropology grew and fieldwork proliferated, scientists periodically conceived classification schemes for different members of the human family that grew steadily in complexity. In 1775, a French scientist named Virey argued that there were two human races. That same year, the German scholar Blumenbach, sometimes called the father of physical anthropology, argued that there were five human races. In 1860, Geoffroy Saint-Hilaire argued that there were only four major races, but that they could be divided into 13 secondary races. In 1870, Thomas Huxley, Darwin's champion, refined this to five major races with 14 distinguishable secondary races. In 1879, the influential Swiss scientist, Ernst Haeckel, argued that the four principal races could be divided into 12 secondary and 34 tertiary races. These various typological schemes were based on systems that measured physical characteristics, assessed language structure, and (in some cases) analyzed cultural patterns.

Despite all the academic arguments, the definition of "race" itself remained elusive. By the close of the 19th century, most anthropologists thought of races as groups that could be delineated by comparing the mean measurements of certain physical traits. One favorite was skull shape (for which there was a cephalic or cranial index); another was nose shape (the metric was for relative flatness, called the platyrhinal or nasal index). Of course, skin color and the shape and coarseness of hair were always considered. Interestingly, as late as 1900, one leading anthropologist, Jean Deniker, chief librarian at the Museum of Natural History in Paris, was still so uncertain about the nature of the differences among humans that he would not exclude the possibility that there was more than one human species. He mused that " No one has ever tried cross-breeding between the Australians and the Lapps, or between the Bushmen and the Patagonians, for example." Deniker was suggesting that such crosses might be infertile, which for biologists is the classic boundary between species. By 1920, few reputable scientists supported the hypothesis that there were more than one human species, but the idea lingered. As recently as 1963, Sir Wilfred Le Gros Clark, a professor of anatomy at Oxford and one of the great names in 20th-century anthropology, observed, "Some zoologists would argue that the different races of mankind are really to be regarded as different species, though it is generally agreed that, by the usual criteria, such a distinction is unwarranted." Throughout the first six decades of the 20th century, the idea that one could use metrics such as the cranial index to define human subspecies (races) remained well entrenched.

Of course, those who specialized in defining racial typology were not without critics. As part of its study of the impact of immigration on the United States, The Immigration Commission in 1907 commissioned the renowned Columbia anthropologist, Franz Boas, to investigate whether one could detect differences in bodily form by comparing measures of immigrants and their offspring who were born in the United States. Boas performed extensive studies of Hebrews (his term) and Sicilians living in New York City. His findings, reported in 1910, led the Immigration Commission to assert that "Children born not more than a few years after the arrival of the immigrant parents in America develop in such a way that they differ in type essentially from their foreign-born parents...even the form of the head, which has always been considered as one of the most permanent hereditary features, undergoes considerable changes." A few lines further, the Commission reflected "May it not be that other characteristics may be as easily modified, and that there may be a rapid assimilation of widely varying nationalities and races to something that may well be called an American type?" Boas's work may reasonably be thought of as marking the beginning of a long, albeit slow and uneven, decline in the relative importance attached to race as a lens through which to understand and define human groups.

The early 20th century witnessed the rebirth of Mendelian genetics and the rise of population genetics. As scientists learned more about genetics, support for a biologically based racial hierarchy among humans declined. The notion of race remained a central tenet of physical anthropology, but as the decades passed, scholars tended to use the term in a more circumscribed manner. The prominent Harvard professor, Edwin Hooton, deplored terms such as white race, Jewish race, Latin race, and Irish race, because the terms suggested that race was a matter of skin color, religious tradition, language, locale, or, perhaps, temperament.

By the 1940s, no doubt heavily influenced by the gruesome apotheosis of racism in the Nazi death camps, some anthropologists were arguing that the concept of "race" itself was intellectually bankrupt. In his 1945 book, *Man's Most Dangerous Myth: The Fallacy of Race*, Ashley Montagu argued that racial typology was one of the worst errors made in the history of human thought. But others remained convinced that there was real value to racial typologies. At mid-century, Carlton Coon, a prominent Harvard anthropologist, still argued for six distinct races that could be further subdivided into 30 "stocks."

One reason for the persistence of racial typologies was that advances in biochemistry provided novel markers, the prevalence of which not infrequently varied by race. Between 1900 and 1902, Karl Landsteiner discovered and characterized the four classical blood groups, O, A, B, and AB. His work laid the intellectual foundation for blood transfusions. During World War I, two Polish army physicians discovered that the distribution of the blood groups varied considerably among "racial groups." When their work was published in 1919, it elicited an enthusiastic response from anthropologists. However, it faded when they found that studies of the ABO group in populations did not always yield results that were compatible with existing distinctions built upon morphological analysis.

As the decades passed, other serological markers (such as the Rh and MN blood groups) were discovered that also differed substantially in their distribution among populations. In 1950, William Boyd, a professor of immunology at Boston University, argued in his book, *Genetics and the Races of Man*, that variations in the gene frequencies which determined these three blood group systems could be used to define six races. A central feature of his argument was that these variations correlated moderately well with geographic distribution of his races. To his credit, Boyd hoped that the use of such neutral measures as blood serology would help lay to rest fallacious arguments about racial inferiority that had been tied for more than a century to skin color.

During the 1960s and 1970s, scientists discovered that just as with blood groups, virtually all enzymes had common variants, the distribution of which might also be used in an attempt to delineate differences between populations. The results of the many such studies that were undertaken delivered a decisive blow against racial typology. The brilliant Harvard population geneticist, Richard Lewontin, was among those who compiled irrefutable evidence that the distribution of variant biochemical markers within any putative human race was almost as wide as the differences in the mean distribution of the same marker between any two putative races. If within-group variation approximated between-group variation, of what value was race?

Hard upon the advances in biochemical genetics in the 1960s and 1970s came the revolution in understanding variation at the level of the DNA molecule in the 1980s and 1990s. One result of our ability to assess even the most subtle differences in the DNA sequences that comprise the human genome is a remarkable paradox. A comparison of DNA samples taken from any two persons on earth (even a Bushman and a Patagonian) will find that they are on average identical in about 999 out of each 1,000 bases that are examined. Yet, these variations ultimately influence many of the physical differences (skin color, stature, nasal shape) that 19th-century anthropologists originally used to categorize races. Even though they constitute only one-tenth of 1% of the genome, because the genome is composed of billions of DNA bases, there are between any two persons (other than identical twins) several million points of variation. Some of these will affect the production or function of proteins that play key roles in cellular life. Whereas some DNA variants will cause small changes in proteins that result in disease, many more (acting in ways too subtle for us to grasp currently) may just slightly alter a threshold of risk for developing one or more disorders later in life.

Now that we are able to examine DNA in exquisite detail, it is possible to ask whether merely knowing some facts about variations at certain locations in the genome would permit one to infer that the sample came from, say, one of the four races (Caucasoid, Negroid, Mongoloid, and Australoid) that doggedly persist in human thinking. Among those who are most excited about this possibility are law enforcement officials, some of whom dream of a day when DNA extracted from evidence found at a crime scene (for example, dried spittle on an old cigarette butt) may be coaxed to declaim that the smoker was a tall Asian male with a history of heart disease. This is not (yet) possible. There is so much variation within groups that the likelihood of being able to infer from analyzing a particular DNA sample that the person from whom it is derived belongs to a particular group is too low to offer practical value. There are gene variants, such as the one that causes sickle cell hemoglobin, that are far more common in African people than in European people. But the sickle cell gene variant is also fairly common among Greeks, Cypriots, Moroccans, and Italians, so the confidence with which one might consistently make a judgment about an individual's race or ethnic group is low. Because about one in ten "Negroid" individuals carries the sickle cell variant, its value as an evidentiary tool is modest.

My *Webster's New Collegiate Dictionary* (9th edition, 1983) defines race as "a division of humankind possessing traits that are transmissible by descent and sufficient to characterize it as a distinct human type." Since all humans possess traits that are transmissible by descent, any test of the definition must focus on the requirement that these traits be "sufficient" to characterize that division of humanity as "distinct." The essence of distinctness is that the group in question can be recognized as a discrete entity. Applying Webster's definition, it may be that assertions about race are actually much more persuasive if they are based on culture and language rather than on physical traits. I know of no group of physical or molecular traits that constitutes a boundary which includes a single group of humans and excludes all others.

The notion of race seems to be slowly passing out of human discourse. One important reason is that there is now incontrovertible evidence that all humans share an ancient hominid lineage which dates back millions of years. It is also beyond dispute that all living humans today share a common ancestor who lived approximately 100,000 years ago. Finally, there is persuasive evidence that today the entire world is populated by descendants of either those who stayed in Africa or those who left it about 55,000 years ago. In a very real sense, everyone in the United States is in fact an African-American!

Despite our several-million-year hominid history, the physical variations that we use to categorize people into races diverged within the last 50,000 years. The vast majority of our evolutionary history transpired before race, as we use that term today, existed. It is beyond dispute that humans are much, much more alike than they are different. Imagine that a team of (peaceful) alien scientists has arrived (cloaked in invisibility) on earth to study its life forms. In preparing their report, "On Humans," I bet they would be much more likely to focus on gender differences (which at times really do seem profound) than on the minor physical differences that define race!

In 1998, the American Anthropological Association issued a position paper in which it asserted that race was a social construct which had been used to rationalize myriad evils ranging from housing discrimination to the holocaust. The organization expressed special concern that racial myths reinforced stereotypes about differences in ability and impeded efforts to disentangle biological and cultural factors in human success and failure. In 2000, on the occasion of the announcement by President Clinton that the first draft of the human genome had been deciphered, two of the geneticists who had led that effort used the occasion to reject any scientific value to the notion of race.

No doubt there will be periods when efforts to tie social problems to race (which happened all too often in the 20th century) will resurge. This happened in 1968 when educational psychologist Arthur Jensen published an article in the *Harvard Educational Review*, arguing that extensive studies showed that early intervention studies such as the Head Start Program had failed to boost school performance, and suggested that this was in part because black persons were on average not the intellectual equals of white persons. The article reignited the century-old debate on race and intelligence, and Jensen was castigated by the liberal academic establishment. In 1994, Harvard psychologist, Richard Hernstein, and, his colleague, Charles Murray, published *The Bell Curve*, a mammoth tome that indicted social programs intended to assist the underprivileged, and which also implied that African-Americans were intellectually inferior to whites. It, too, evoked a withering counterattack.

If we review the various efforts to identify major racial groups over the last 250 years, the overwhelmingly most important variable seems to be skin color. In the 19th century, the United States really was a nation of white people, black people, red people, and yellow people. In 1892, a Harvard professor assessed the physical traits of a freshman class from which he created a composite Harvard student. The composite was a white male, about 5 feet, 10 inches tall, with the elongated head shape (dolichocephaly) historically associated with peoples of northern Europe. Imagine repeating this quaint exercise today. Of course, there would have to be a composite man and a composite woman. The composite's height would probably be a bit shorter, despite the overall increase in human size during the 20th century. This would reflect the fact that Harvard now has many students whose ancestors lived in parts of the world where folks tend not to be tall. The skin color would probably be a light brown, reflecting the contributions of the African-Americans, Latinos, Asians, Polynesians, and others who make up about a third of the class.

As anyone who has ever filled out a census form knows, racial categorization continues to reflect and influence governmental policies. One current example of this is that hundreds of Americans who number both Seminole Indians and black slaves among their ancestors are struggling with the Federal Bureau of Indian Affairs to win government benefits that "*non-black*" Seminoles currently enjoy. In an ironic twist of fate, the leaders of the Seminoles have argued that people of African *and* Seminole heritage are insufficiently Seminole to be counted as true tribal members. The Bureau of Indian Affairs has tried to avoid the issue, granting some benefits to black Seminoles, but falling far short of giving them parity with less admixed Seminoles. The Seminole tribe, which suffered much racial discrimination over the last 150 years, appears to have adopted some of the lessons of prejudice.

Ironically, genetics, the science that debunked racial typologies based on crude phenotypic factors, is providing new knowledge that (unintentionally) shores up arguments to maintain them. A growing body of evidence suggests that racial disparities in health and disease are caused not only by socioeconomic factors (albeit immensely important), but also manifest due to differences in the distribution of genetic variants. Consider the recent finding that certain widely used blood pressure medicines called ACE inhibitors are less effective in blacks than in whites. Recently, a drug called eplerenone has been shown to be more effective in lowering blood pressure in blacks than in whites. The reason for this probably is that there are gene variants more common in one race than in the other that produce slightly different enzymes which respond differently to the drugs in question. In some instances, race (a surrogate for some subtle physiological variation) may dictate choice of therapy when a doctor is trying to treat high blood pressure. The field of pharmacogenetics—in essence choosing drugs on the basis of genetic tests of the patient—could sustain racial typology, but hopefully, for constructive purposes. Over time as pharmacogenetics matures, DNA sequences will be of much greater value than skin color in guiding choice of medication.

Perhaps the most persuasive evidence that the concept of race will remain important in health care is the commitment made by the National Institutes of Health to the HapMap project. After collecting DNA samples from people in China, Japan, Nigeria, and the United States, scientists are studying thousands of spots in the genome to look for common patterns of variation (two or more of these comprise a haplotype). Eventually, they will construct a book of haplotypes. Each page in this book will be composed of a block of DNA that represents a common variant which has remained stable over the generations. They are likely to find many haplotypes that are associated with an increased risk of disease or with a positive response to a particular therapy.

In 2003, Essie Mae Washington-Williams, a 77-year-old retired schoolteacher, made headlines when she disclosed that she was the daughter of South Carolina Senator Strom Thurmond, for decades one of the nation's most zealous segregationists. Her mother, Carrie Butler, a maid in the Thurmond home, was 16 years old when she gave birth to Essie Mae. On July 1, 2004 her name was added to the monument to Strom Thurmond on the statehouse grounds in Columbia, joining the names of his other, white, children. Ms. Washington-Williams continues to make headlines. She announced that she wanted to join the United Daughters of the Confederacy, an almost exclusively white organization. If that organization accepts her application, as it has said it will, this event might be remembered as a milestone in the collapse of racial barriers in the United States. Of course, it will pale before Barack Obama's history-making run for the Democratic presidential nomination.

One of the most important discoveries flowing out of molecular biology in the last quarter century—possibly, the most important—is that humans are remarkably similar. If one randomly selects two persons of any gender, race, or ethnic background from anywhere on earth and performs a comparative analysis of their DNA sequences, the two will be 99.9% alike. The irrefutable truth is that the typology of race is based on relatively unimportant physical characteristics that emerged under environmental pressures during the last 50,000 years. Wherever you travel on the face of this small planet, there is a chance that your genetic makeup qualifies you as a potential donor of a life-saving kidney to the last stranger you noticed. Of course, the same is true for the stranger.

Two elderly Salishan women, Wyl-lehy, age 110, and her niece Kitty, age 70. Circa 1904. (Photo by Bartholomew Danihy, Courtesy of the Library of Congress.)

Longevity

Have you heard of the wonderful one-hoss shay,
That was built in such a logical way
It ran a hundred years to a day,
And then of a sudden, it —ah, but stay,
I'll tell you what happened without delay...

O.W. Holmes, *The Deacon's Masterpiece*, 1858

In the spring of 2004, Aurelia Marotta, age 113, of East Boston, who had been pretty lucky most of her life, had an unlucky death. If she had lived just 26 hours longer she could have claimed the title of the oldest person in the United States, for the only person older than she, Elena Slough, died the next day. Except for her near miss in the Guinness Book of Records, Mrs. Marotta seems to have had a fine life. At 100 she routinely went grocery shopping, pushing her own cart through the aisles. She lived at home until age 108. According to her children, she had a fine memory to the end of her life. In the obituary, they wrote that their mother never offered an explanation for her extraordinarily long run of good health.

I have been collecting the obituaries of centenarians (admittedly, an odd hobby) for the past few years, partly in the hope of finding a secret thread that connects their otherwise disparate lives. Except for the facts that virtually none of them ever smoked cigarettes for long, that they tend to be of small stature, and that they generally have upbeat, outgoing personalities, I have not found many clues. Sometimes, I take out my obituary collection and reread it. Often, I wind up thinking that one or another of the centenarian club is among the luckiest people ever to grace the planet.

When I am in a contemplative mood, I envy Alexander Skutch, who died at 99. Skutch, whom I include in my collection because he lived independently to the end, was undoubtedly the world's authority on the birds of Central America. He lived for 60 years on a beautiful hillside in Costa Rica, achieving a rare oneness with nature. When I am in a more gregarious mood, I think of James Stillman Rockefeller. Born into one of America's most fabled families (he was the grandnephew of John D. Rockefeller, the founder of Standard Oil), James graduated Phi Beta Kappa from Yale. In 1924, he was captain of an eight-oar boat that won a gold medal at the Olympic games. Over the next three decades, he led a distinguished career in finance, rising to become chairman of First National City Bank of New York. In World War II, he served with distinction in the Airborne Command, achieving the rank of lieutenant colonel. He was a lifelong philanthropist who provided substantial support to many New York City institutions. He had a solid marriage that lasted 68 years, producing four children, 14 grandchildren, 37 great-grandchildren, and one great-great-grandchild! He was in good health until the day he died at the age of 102 in 2004. It does not seem fair.

Perhaps the oddest member of my obituary collection is Max Schmeling, who during the 1930s was one of the world's most famous men. In June 1936, before a huge crowd in Yankee Stadium, Schmeling, a German, thought by virtually everyone who followed boxing to be the underdog, knocked out unbeaten Joe Louis, the black American champion, in the twelfth round. Hitler was delighted and ordered films of the fight to be shown throughout Germany. The Nazi propaganda machine used the victory as proof of Aryan (white) supremacy. Two years later, in a rematch that would determine who was the world heavyweight champion, Louis broke two of Schmeling's vertebrae (with a legal kidney punch) in the first round. The fight lasted 124 seconds. The Nazis did not show the fight film. Given the extremely low likelihood that anyone born in 1905 would live to 2005, how can it be that a professional boxer, who in scores of fights took innumerable, violent blows to his head, could do so? In his old age, Schmeling did not even suffer from serious dementia, which so often affects boxers in their fifties. I have no explanation for his longevity, but I do hope he donated some DNA for research.

My favorite centenarian is Ernst Mayr (also a German), whose obituary ran in the *New York Times* on the morning I was writing a draft of this chapter. Mayr, the greatest evolutionary biologist of the 20th century (often called its Charles Darwin) began professional life as an ornithologist who specialized in the birds of New Guinea. He described 24 species of birds for the first time, a feat unequaled in his lifetime (he also delineated nearly 400 subspecies). Mayr posited and proved that new species arose out of geographical isolation. Even greater was his effort during the 1940s and 1950s to integrate evolutionary theory with the new science of molecular genetics, an intellectual undertaking now remembered as the "modern synthesis." He contributed over 600 scientific papers. He published the last of his 20 books when he was one month past his 100th birthday! I had the privilege of meeting Mayr on several occasions, and I know several people who knew him well. Like many centenarians, he was a diminutive person, but, unlike most, he was almost uniformly regarded as an irascible individual.

Although there is no chance I will ever be born into money, be elected to Phi Beta Kappa at Yale, or win an Olympic medal of any kind, I still have a (small) chance to make it to 100. The fastest-growing human cohort in the world is made up of the extremely old. Between 1950 and 2000, the chance of living to be 100 increased 20-fold. When pressed, scientists offer a litany of reasons. The most common are the benefits of vaccines to protect against once commonly fatal disorders of childhood, much cleaner air and water, powerful antibiotics to fight infections like bacterial pneumonia, advances in occupational safety, awareness of the dangers of smoking, treatment of high blood pressure, and much improved care for heart disease.

Admittedly, these answers are not much different from those that would have been proffered 30 years ago. However, when you are thinking about changes that affect human longevity you have to use a long time line. In any given year, there is at least one development that can logically be expected to improve the overall odds of living to 100. A recent example is the impact of population-wide vaccination against chickenpox (varicella). After comparing death records from the five years prior to the introduction of the vaccine to records during the first few years of its use, epidemiologists concluded that the vaccine has already saved the lives of about 100 children. Most of them will live to a robust old age; one or more may make it to 100.

The Molecular Biology of Longevity

As a walk through virtually any old graveyard in New England (another of my favorite pastimes) will prove, it is nothing new to live into the 90s or to reach 100. A highly unscientific impression I have formed from reading gravestones is that some families are long-lived. During the last decade, gerontologists and geneticists have marshaled impressive evidence that the genetic cards one drew are at least as important as many of the recognized environmental factors in limiting or promoting longevity. Indeed, a few gene variants may turn out to be critically important.

Thanks to studies of model organisms, we are on the brink of understanding how genes influence life span. Over the last 15 years, scientists have made extraordinary advances in identifying key genes and key environmental forces that interact to set the outer limits of longevity. With these advances, the idea of someday applying that knowledge to extend the human life span has moved from science fiction to a scientific (albeit still remote) possibility. Understanding biological limits is a formidable challenge, and many people have contributed to the current state of knowledge about life span.

One of the most important historical threads starts on January 13, 1927, the day Sydney Brenner was born to poor Jewish immigrants in Germiston, South Africa. A precocious young man, Brenner enrolled in the University of Witswatersrand in Johannesburg before his 15th birthday. Over the next decade, he established himself as a brilliant young researcher and an indifferent medical student. By 1953, Brenner held a junior appointment at Oxford University. Already fascinated with DNA, he visited Watson and Crick in Cambridge just days after they had elucidated its double-helical structure. During the 1950s and 1960s, Brenner, with a score or so of other scientists, created the field of molecular biology. He spent 20 years (1956–1976) sharing an office with Francis Crick at the Cavendish Laboratory in Cambridge.

By 1963, Brenner, one of the great theoreticians of 20th-century science, was convinced that the major problems in molecular genetics were solved or would soon be solved. In a letter written in June of 1963 to Max Perutz, his departmental director, Brenner argued that the real challenges lay in two other fields, developmental biology and the organization of the nervous system. He wrote that he was thinking of a new approach to understanding how organisms develop from a single cell, and that he "would like to tame a small metazoan organism to study development directly." Within a few months he had chosen the organism that he would study for the next four decades, a tiny worm called *Caenorhabditis elegans* (*C. elegans*).

C. elegans is little over a millimeter long. It lives in the wet spaces between damp soil particles or on decomposing leaves, eating mostly bacteria. It moves in an undulating manner, reflexively drawn to favorable environments and reflexively avoiding unfavorable ones (dry, hot spaces). Composed of only about 1000 cells, the little hermaphroditic worm nevertheless possesses a hydrostatic skeleton (an elastic membrane that maintains an outer coat rigidity using high internal hydrostatic pressure), a pharynx, a digestive tract, muscles, and germ cells. It bears about 300 offspring at a time. One major reason Brenner found this little worm so attractive for research is that its offspring are clones (genetically identical).

In 1964, Brenner, joined by a steadily growing number of top students and colleagues, set out to learn everything he could about the life plan of *C. elegans.* Despite his fame, in the first years, the project was not well publicized. But in 1974, he published a seminal work on the genetics of *C. elegans* that put the organism on the research map. Over the ensuing 30 years, scientists have worked out the fate map of the little worm, tracing how *every* cell that is ever part of the worm comes into being, moves to its proper place, functions, and dies. When genome sequencing matured in the 1990s, *C. elegans* was the first animal genome to be completely described base pair by base pair. In 2002, Brenner and two colleagues, John Sulston and Robert Horwitz, shared the Nobel Prize for their part in realizing Brenner's dream of completely understanding the development of every cell in a multicellular organism. One anecdote claims that in Sulston's office at Cambridge there are grooves in the wooden floor caused by the thousands of trips he made in his chair from microscope to writing desk and back as he painstakingly recorded the life history of each cell.

By the late 1970s, the life history of *C. elegans* was so well understood that it drew the attention of a small group of scientists interested in the genetic control of aging. In 1988, two biologists in California made the electrifying report that they had found a mutation in a *C. elegans* gene that extended the worms' mean life span by 40% and the maximum life span from 22 days to 46 days (a very long time if you are a nematode!). They named this first genetic locus in which a mutant form lengthens life, "age-1." Their paper stimulated studies of the molecular genetics of aging. Within a decade, it became a major field in modern biology.

Since about 1993, Cynthia Kenyon, a scientist at the University of California at San Francisco, has moved the study of the genetics of longevity rapidly forward. In that year she reported the discovery that certain muta-tions in a single gene called daf-2, which works in tandem with a second gene called daf-16, can cause fertile, active adult worms to live twice as long as worms lacking the mutations! Over the ensuing decade, Kenyon and her colleagues have published a flood of papers that are redefining our thoughts on life span. These include the discovery that if the cells that give rise to germ cells are killed (via laser beam), the sterile worms live *much* longer, that certain mutations in genes that control electron transport (en-ergy transfer) extend life, and that prolonged caloric restriction greatly ex-tends mean life span. When Kenyon and her team killed the germ cells in worms that had the daf-2 mutation, the animals lived six times longer than their normal counterparts (140 days compared to about 22). Throughout this vastly extended period, the long-lived worms remained as active as normal adults! They were both ancient and robust! Kenyon has taught us that (at least in worms) the requirements for a long, healthy life are the right mutation, no reproduction, and very little food—not an especially attractive lifestyle.

For nearly a century, the workhorse favored by genetic researchers has been the fruit fly (*Drosophila melanogaster*). Success in finding aging genes in *C. elegans* aroused the "fly community." Beginning in the 1980s, Michael Rose, an evolutionary biologist at the University of California (Irvine) has laboriously searched for mutations that promote longevity in the fruit fly. Merely by selecting the most long-lived flies in laboratory populations and using them as progenitors of a new cohort, Rose has been able to develop strains that live twice as long as their ancestors. He, too, has found that by restricting reproductive activity and limiting access to calories, he can even further extend the life expectancy of his flies. Of special interest to the mil-lions of people who consume nutritional supplements purported to be an-tioxidants, Rose has shown that flies with certain mutations in the gene for superoxide dismutase, which codes for a protein that scavenges free radi-cals, live much longer than the wild type. Others have shown that flies with mutations in the gene that makes an enzyme involved in the regulation of brain signaling, called dopamine decarboxylase, are also long-lived. One group asserts that this gene alone accounts for 15% of the differences in life span.

Despite the vast differences that separate them on the evolutionary tree, mammals appear to be under the same genetic and environmental influences as flies and worms. In 1935, Clive McKay, a nutritionist working at Cornell University, discovered that he could extend the life span of rats. Merely by cutting their average daily caloric intake by 30%, McKay routinely extended their lives by about 40%. Although such experiments have only been done in a few mammals, lifetime dieting seems to confer longer life span on all of them. A few years ago, researchers at Cornell's Baker Institute for Animal Research showed that caloric restriction markedly extended the lives of dogs. In one experiment, almost all of the calorie-restricted dogs outlived all of the dogs given regular diets. Dr. Doug Antzcak, director of the Baker Institute, told me that the more long-lived dogs did not seem to enjoy life nearly as well as did the better-fed ones. Studies of cows, guinea pigs, and rhesus monkeys also support the positive impact of persistent calorie restriction on life span. It is hard to believe that many humans would trade a few years of extreme old age for a lifetime of near starvation. There must be a better way.

Yoda, a diminutive fellow who lived in a rest home for geriatric laboratory mice run by Richard Miller, a pathologist at the University of Michigan, garnered his 15 seconds of fame in April of 2004 when he celebrated his 4th birthday. The average laboratory mouse lives just 2 years. Yoda's longevity was due to genetic modifications that altered the function of his pituitary and thyroid glands, and reduced his baseline production of insulin. These genetic alterations allowed Yoda to double his anticipated life span without the need to live a calorie-restricted diet—the first such mouse to do so! Unfortunately, Yoda died just twelve days after celebrating his 4th birthday. The press release that the University of Michigan issued about his demise (possibly, the first mouse obituary in history), quoted researchers as saying that Yoda had lived a life span equivalent to 136 human years.

Longevity in mice is a lot more interesting than longevity in worms and flies. Mice are a lot like humans. Humans have intensively studied the genetics of mice for nearly a century, and it has long been known that some inbred strains live much longer (or shorter) than average. Scientists have shown that life span correlates well with average timing of cell division (cell cycling time). In 1999, mouse geneticists discovered that by disrupting the function of a single gene (p66) that is involved in programmed cell death, they could increase mouse life expectancy by 30%. Yoda's longevity is attributed to the fact that he is a dwarf due to a pair of recessive genes that also slow his metabolism. Among mammals (including dogs and humans) it is generally the case that small individuals live longer than big ones.

One exciting recent discovery in the field of longevity was made by a Boston researcher, Leonard Guarente. For the last decade, Guarente has been searching for gene variants that mimic the positive effect on life span achieved by rigorous calorie restriction. In 1999, he found a gene in yeast called SIR2 that seemed to have that effect. If he inserted extra copies of SIR2 in yeast, they lived markedly longer than regular yeast; if he deleted the gene, the yeast died young. In 2004, he and his colleagues showed that the mouse version of the same gene (called SIRT1) also substantially influenced longevity. In addition, they began to understand the mechanism. The protein made by SIRT1 turns off genes which make proteins that help to store fat. It also commands liver cells to break down fat. These two effects keep the mice lean and long-lived! These discoveries led Guarente to found a company that is trying to develop compounds which regulate SIRT1 in mice and humans.

GENETICS AND HUMAN LONGEVITY

Interest in the human life span is as old as history itself. Genesis (Chapter 5) tells of Adam living 930 years, Seth 912, Enoch 905, Methuselah 969, and Noah 950 years. Such ages are scientifically incomprehensible. But what of Abraham who lived 175 years and "died in a good old age" (according to the Revised Standard Version), or Ishmael who lived to be 137? Might these ages be attainable in the modern world? Currently, the record is held by a French woman, named Jeanne Calmet, who lived to 122. She was one of the very small group of super-centenarians—those who reach 110. Right now in the United States, there are a relatively large number of people aged 114, but none who are 115. This could be due to chance, or it could reflect some biological barrier we have yet to understand. Still, I am willing to bet that in the 21st century a human will reach the age of 137.

In so saying, I line up with Steven Austad, a researcher at the University of Idaho, who has made a fascinating wager about the human life span with a colleague named Jay Olshansky. Each man has bet $150 on whether or not a human being will live to age 150 before January 1, 2150. Olshanksy says no; Austad is confident it will happen. That is, he believes there is someone alive *today* who will be alive in 2150 and who will be sufficiently intact to appreciate his or (more likely) her environment. Austad argues that stem cell research will lead to new anti-aging technologies that will make it common for people to live well beyond 100. Over the last few years, his views have attracted many supporters. Some demographers now think that a baby born in the United States in 2060 will have a 50/50 chance of becoming a centenarian! Applying historically reasonable compound interest rates, the jackpot in the Austad–Olshansky wager should balloon to hundreds of millions of dollars by the day of reckoning! If neither man has living direct descendants, the fortune will go to universities.

What do we know about the genetic influences on human life span? The candid answer is "very little," mainly because humans are exceedingly difficult to study. It would be immoral (and illegal) to alter human embryos to determine how a changed gene affected life expectancy. Furthermore, we are already so long-lived that any scientist who hopes to determine whether interventions affect life span will (like those who designed the great cathedrals of the late middle ages) have to be content knowing that persons not yet born will analyze the results. We do not even have any biomarkers that we could monitor to discern a trend which is suggestive of increased life span.

Until recently, we were not even confident about the limits of the current human life span. There are, not surprisingly, all sorts of problems documenting the accuracy of birth dates for people born even as recently as the late 19th century, so claims of extraordinary age are often suspect. Fortunately, beginning about the turn of the 20th century in Europe, efforts to record human births and deaths improved substantially, so we are entering a period in which it will be much easier to verify age among the very old.

Although humans have probably always wondered what the upper limits were on their existence, it was not until the 1920s that scientists began to consider this question in a formal way. Raymond Pearl, a geneticist at Johns Hopkins University, studied a large cohort of families noted for longevity and concluded that parental age at death was a relatively good predictor for life span of offspring. However, further studies of the predictive power of parental age at death suggest that it is of modest value to offspring. Pearl developed a theory of aging called the rate of living theory. He posited that each organism is born with a sort of energy bank, and that the more energy expended per weight over time, the shorter the life span.

The modern version of Pearl's thesis is that the rate of cellular energy metabolism (largely determined by mitochondrial function) places broad limits on the life span.

One of the earliest studies of the genetics of human aging was the New York State Psychiatric Institute Study of Aging Twins that was initiated in 1946. Directed by Franz Kallman and Gerhard Sadler, the study focused on 1603 twin "index cases" that had reached a well-documented age of 60. Over time, a subset of 268 men, including both identical and fraternal twin pairs, has been followed to extreme old age or death. The data demonstrate that among identical twins the intra-pair differences in life span are much smaller than among fraternal twins. In effect, if you are an identical twin, the death of your co-twin is an ominous sign!

As our ability to sift through the human genome has improved, mining for longevity genes has begun to yield good-quality ore. In 2001, a Boston-based group of scientists reported that their search for genetic markers that are associated with having reached extreme old age had discovered a region on chromosome 4 that appears to harbor a longevity allele. In 2003, a team of scientists at Elixir Pharmaceuticals, Inc., which studies the lives and DNA of centenarians around the world, reported that a variant of a gene for microsomal transfer protein was strongly associated with increased life expectancy. In 2003, another group reported that in a study of older Ashkenazi Jews, a variant of a gene involved in cholesterol transport was much more prevalent among 200 persons with a median age of 98 than among a control group of 70-year-olds!

On the basis of the studies in model organisms, we can draw a number of inferences about the genetics of life span in humans. One widely accepted hypothesis is that a gene called insulin-like growth factor-1 (IGF-1) is a key player in an evolutionarily derived regulatory system that regulates animal aging as a function of food availability. In addition, a number of other genes affect life span in a less direct manner. If we incorporate what we know about environmental factors, we can assert: (1) life span is under genetic control, (2) many genes are involved and we will soon know all of them, (3) some genes are more influential (possibly, much more) than others, (4) to maximize the odds of reaching the genetic limit of human life requires strict attention to environmental risk factors, (5) long-term caloric restriction is for now essential, (6) it may be possible to develop drugs that target longevity genes in a manner that permits one to circumvent the demands of caloric restriction and other unpleasant activities, and (7) ultimately, it will be possible to use genetic engineering to enhance human embryos with longevity alleles.

If powerful life extension strategies are developed, they will raise some really disturbing ethical and public policy questions. For example, since it is quite unlikely that either government or private insurers will pay for life extension therapies, they will (like other technologies such as infertility treatments) be accessible only to wealthy members of society. One could argue that this fact is congruent with current patterns in the consumption of health care. If they are willing to consider surgery outside the United States, wealthy people currently can obtain needed kidney or heart transplants in some other nations without spending months on a donor waiting list. To put it more cynically, they can jump to the head of the line.

The biggest dividend from understanding the genetics of longevity will be our discovery of genes that fight senescence. Without question, genes shape the likelihood that we will live in good health into our 90s. Almost certainly, genes that regulate cholesterol metabolism, lung function, the immune system, mood, and a myriad of other aspects of our lives define whether we will make it to 90 and in what condition. What we will aspire to do is figure out which variations in which genes increase the chances that we will live robustly for 100 years or more. After finding them, we will need to learn how to modify their action both to help those poor souls born without the preferred variations and to improve further the benefits conferred by the good variants. Ultimately, like the one "hoss" shay depicted by O.W. Holmes in his poem, *The Deacon's Masterpiece*, we might deduce how to end a long, robust life with a built-in very rapid demise.

Louis M. Terman, the Stanford professor who made IQ a household word.

Intelligence

We live in a world described by bell-shaped curves. Most things that humans measure in large samples distribute normally around a mean. If I were to superimpose graphs of the distribution among a large cohort of individuals of measures of height, weight, body temperature, the age in weeks at which infants took their first unassisted steps, age at death, the lifetime batting averages of all major league baseball players, the times it took schoolboys of equal age to run a mile, or innumerable other metrics, the data points would define a bell-shaped curve with 95% of them falling within two standard deviations of a vertical line running right through the apex of the curve. At each end of the curve, I would find the outliers—those with measurements most different from the mean value.

Humankind loves to think about the outliers, especially those who occupy the extreme right side of the tail of the bell curve. Here, if we were so inclined to review the data, we would find the tallest, oldest, smartest, and richest people. Here, too, are the sub-four-minute milers, the Hall of Famers, the Olympic medalists, the child prodigies in chess and music, and the high school juniors who scored 1600 on the SAT I. For any given metric, each of us occupies a distinct point on an extremely large number of possible bell curves. For most metrics, we fall within the two standard deviations from the mean. For one or more, we may be way out on one of the tails (hopefully, on the right side!). Those who occupy a point three or more standard deviations to the right may, if the metric is valued, be lauded and envied by others.

Even if the distribution in question describes a mundane metric, such as height, the outliers fascinate. Who has not stopped to stare when a man who is over seven feet tall walks by? A few years ago, I was at an exhibition at the Boston Museum of Fine Arts. In the tightly packed space, the seven-foot-tall Chinese man who was intently studying the art garnered as much attention from the sophisticated crowd as did the paintings. Of course, height is (unless tied to some other ability) a mere curiosity. We are far more interested in hearing the great musicians play, watching the top athletes run and swim, following the exploits of teenage chess prodigies, and contemplating the paintings of the great artists. Humans have always been fascinated by genius. Since ancient times, those of us trapped near the mean have wondered how extremely talented people achieve that status. The suspicion that unusual talent such as intellectual genius is inborn is tied to the word itself—genius derives from the Greek, referring to a condition at birth.

The word entered the English language during the Renaissance, originally meaning interests or proclivities that one has from birth. Long before it was used to describe persons of "transcendent mental superiority" or "a person associated with a very high intelligence quotient" (two definitions in my Webster's *Ninth New College Dictionary*), genius was used to describe persons of outstanding creativity in the arts.

The first person to conduct formal studies on the influence of heredity on the expression of something akin to genius was the Victorian polymath, Francis Galton. No doubt influenced by the impact of Charles Darwin's (his cousin) magisterial work, *The Origin of Species* (1859), in the early 1860s Galton began to investigate whether he could find unexpected concentrations of accomplishment within families. Given the cultural constraints of the day, he confined his studies to English men. His early work on the judges of England between 1660 and 1865 was first published in the popular *MacMillan's* magazine in 1865.

His book, *Hereditary Genius* (1869), summarizes detailed studies of eminent statesmen, members of the peerage, military commanders, writers, scientists, poets, musicians, painters, clergy, senior classics (top students) at Cambridge University, and nationally ranked oarsmen and wrestlers. His research method was simple. Galton collected extensive biographical and family histories on cohorts of men in each of the above groups and investigated whether they had ancestors who reached similar pinnacles of achievement. Today, we might say that he studied people who for some valued measure deviated at least two standard deviations from the mean, and then asked whether an unexpected number of their relatives could be said to have reached similar levels of accomplishment.

Galton hypothesized that for every category examined, he would find an unexpectedly large number of relatives whose achievements also placed them on the far right of some distribution. He recognized that wealth, family connections, and other environmental forces could influence life achievements, but he was so thoroughly convinced that much talent was inborn that he did not trouble himself with methodological flaws which epidemiologists would today dismiss as hopelessly biasing the inquiry. It is not at all surprising that Galton's magazine articles and books enjoyed a wide readership. It must have been quite comforting to the ruling class to be told that the eminence which many of their members had achieved was highly heritable. It helped to justify the class structure and provided an argument for its perpetuation.

In studying the role of heredity in academic accomplishments, Galton analyzed the lists published by Cambridge University of the men who had scored highest in the exams called the Classical Tripos in the years from 1824 (when the rankings were initiated) through 1869. In 36 years of the 46-year span, a single man won top honors; in the other 10 years, two men shared the honor. Galton noted that men with the surname Kennedy had taken the first on 4 of the 46 years and that a Lushington family member had done so twice. From his study of the families that produced the independent winners, Galton found that most included several relatives who had led distinguished careers (which he defined as having achieved recognition at the national level). In Galton's view, the evidence suggested that hereditary factors in these families strongly predisposed to high intelligence, top scholastic achievement, and subsequent success in life.

Galton, who coined the word "eugenics" in 1883, had a major influence on the shape of eugenical thinking in England from 1869 through his death in 1905, and beyond. At death, he made a major bequest to the University of London to enable it to continue research in eugenics, largely along a plan he had developed! Unlike the United States and much of the rest of Europe (which focused on social policies to restrict reproduction by the unfit), in England, the Progressives (following Galton's views) devoted their efforts to encouraging marriage and childbearing by talented couples. Today, such a social policy may seem quaint or even silly, but it was much discussed in its day, and it is still espoused in some quarters. During the 1990s, Singapore introduced social programs to encourage marriage and childbearing by well-educated couples. In Europe, where many nations currently are experiencing such sustained low birth rates that they anticipate sharp declines in population and fear the cultural impact of immigrants, pro-natalist policies (which carry an unspoken assumption about superior bloodlines) are much discussed.

In the United States, eventually under the leadership of researchers associated with the Eugenics Record Office at Cold Spring Harbor Laboratory, the major approach to family studies was to document the social costs generated by persons of low intelligence. From the publication of *The Jukes* (1877) until about 1930, more than a score of book-length studies of degenerate families frightened American readers and helped set the stage for laws limiting the right to marry, permitting the sterilization of retarded persons, and restricting immigration by persons with any of a long list of health problems. Perhaps the most famous of this genre was *The Kallikaks* (1912), a study of a rural New Jersey family that was not infrequently summarized in the pages of our nation's high school biology texts.

In the midst of a growing interest in the influence of heredity on intelligence, America discovered the IQ test. Its inventor, Alfred Binet, director of the psychology laboratory at the Sorbonne in Paris, grew up in an era when studies of the human brain were primitive. Phrenology, the "science" of inferring cognitive skills, personality traits, and moral attributes from the contours of the skull, dominated mid-19th-century medicine. The only quantitative studies in neurology sought to correlate brain volume with intellectual capacity. Craniometry focused on between-group differences and was, thus, closely tied to ideas about racial classification that so occupied anthropologists. About 1898, Binet thought of developing simple tests which would correlate well with the widespread assumption that skull volume was a good predictor of intelligence. After three years of research, however, he became convinced that (except for extremely small brains associated with some kinds of mental retardation) skull size did not correlate with intelligence.

In 1904, Binet was commissioned by Parisian educators to develop a test to screen school children to identify who would benefit from alternative education. In response, he developed a series of questions scaled by increasing difficulty. Naturally, the respondents generated scores that were distributed normally. Binet was able to show that the performance scores were highly reproducible, making it relatively easy to arbitrarily decide that the students who scored in the lowest 3% were the ones who needed special help. In 1912, Stern, a German psychologist, took the simple step of creating the IQ scale by dividing the "mental age" by the child's actual age and converting it to base 100. Thus, a 10-year-old who scored at the level of the typical 12-year-old was said to have an IQ of 120.

At the same time that Binet was developing IQ testing, Charles Spearman, a British engineer turned psychologist, was attempting to develop a more sophisticated approach to understanding and measuring intelligence. In 1904, when he was a graduate student in Leipzig, Spearman introduced the concept of "g," a quantitative measure that he argued represented general intelligence. In his famous paper, "General Intelligence Objectively Determined and Measured," he argued that the score on all tests to measure intelligence reflected two kinds of variables (factors), one or more unique to the specific test and a general factor that somehow influenced the scores generated on all the sub-tests. The most sustained argument in the field of psychology, now a century old, is whether Spearman's "g" reflects an underlying intellectual capacity and, if so, the extent to which it is heritable. The pro-Spearman camp asserts that "g" is a measure of brain plasticity or "processing time" and that its heritability is high (0.6 to 0.8 on a scale where 1.0 indicates complete heritability), whereas opponents argue that we have no idea what "g" reflects and, anyway, its heritability is low (0.3 to 0.4). Imported from France to the United States by H.H. Goddard, the psychologist who wrote *The Kallikaks*, IQ testing spread rapidly. By 1917, it was in routine use by the military to evaluate inductees. The advent of what was then widely accepted as a useful tool to measure intelligence (e.g., to plot a person's place on its bell curve) set the stage for the most influential study of intelligence ever conducted in the United States.

A psychologist named Lewis M. Terman made the IQ test a household word. The eighth child of an Indiana farmer, Terman was born in 1877, the year that *The Jukes* was published. He was precocious and ambitious. At 17, Terman was teaching in rural schools, but a few years later he abandoned a job as a school principal to earn his Ph.D. at Clark University in Worcester, Massachusetts, then a major academic center in psychology. Terman's 1905 thesis, a study of the intellectual abilities of seven bright and seven dull boys for which he developed cognitive tests, presaged his life's work. He moved west, and, after a stint as a psychology professor at UCLA, he joined the Stanford faculty in 1910, where he remained until his retirement.

During World War I, Terman, an army major, served on the committee that set up IQ testing to screen recruits. Heavily influenced by his experience with this screening tool during World War I, in 1921, Terman, along with his student, Catherine Cox, initiated a study of gifted California school children, a population that they closely followed from 1921 through 1946. Terman conducted this extensive research in an attempt to combat what he perceived to be a widespread tendency in society to devalue really bright students. The work, which led to many papers and four books, detailed the life histories of 1,528 California school children with IQs above 140 (a score achieved by no more than one percent of those tested), including 300 with IQs above 170. He and his colleagues assiduously followed these kids (nicknamed "Termites" by the popular press) until they reached mid-life.

Terman's research dispelled two long-standing myths: that intellectually gifted persons were psychologically abnormal and physically fragile, and that they tended to lose their cognitive powers in adulthood. As far back as 50 A.D., the Roman historian, Seneca, observed that there is no great genius without a touch of madness. The 17th-century poet, John Dryden, wrote, "Great wits are sure to madness near allied, and thin partitions do their bounds divide." To the contrary, Terman found that the Termites had neither a greater burden of psychiatric problems, nor more health problems, than similarly aged persons of average intelligence. He also found that many (by no means all) of his subjects excelled at university and in their chosen fields. His research gave a certain social cachet to intelligence and did much to legitimize the use of the IQ test as a tool for early identification of bright students.

During the 1950s, IQ testing was widely used in the public school systems and became a portal for entry into advanced programs. It is not merely coincidental that the 1950s saw the advent of popular television game shows like "The $64,000 Question." Despite the title of his project, "Genetic Studies of Genius," Terman did not study the intelligence of the parents of the gifted children, nor did he study their offspring. He did, however, study the occupational history of the fathers, finding that one-third of them were members of the professional classes. Because this is significantly higher than the percentage of parents of school children with average IQ scores who worked in the professional classes, Terman viewed occupational class as a marker for intellectual capacity.

The debate over the relative role of genes and environments in intellectual or artistic achievements breaks down when the focus shifts from analyzing the broad range of achievements within populations to examining individuals who indisputably exceed the capacity of all but a few people on earth. A favorite pastime of those who write popular works about intelligence is to infer the IQ scores of men and women of world renown, such as Goethe, Kant, Voltaire, Newton, Mozart, Darwin, Curie, and Lincoln. Estimates are typically in the range of 170–200. This silly, yet engaging game, misleads the audience into thinking that persons of high IQ lead extraordinary lives. One finding of the Terman study is that—with the exception of a couple of physicists—by age 35, very few of the subjects with the highest scores (IQs above 170) had accomplished feats that would be considered world-class. This is not surprising. High as it sounds, about 1 in 1000 people scores an IQ of 180. In contrast, far fewer people earn a permanent niche in the pantheon of history. And many of them (I think now of presidents and generals) unquestionably have IQs that are unremarkable. Clearly, much more than a high IQ mediates such achievements.

The use and influence of IQ testing in our society has been in a slow decline since the 1960s. The various IQ tests have been subjected to a withering barrage of criticism, especially that they are culturally biased to give an advantage to middle-class white children over black and Hispanic children. Of course, as anyone who has applied to college knows, testing to measure intellectual ability remains alive and well. The SAT I test is probably the most influential cognitive test deployed in the history of civilization. Although little discussed in the popular press, SAT I scores correlate closely with IQ score.

The revolution in our ability to study the DNA molecule has provided researchers with a new approach to dissecting the genetic contribution to intelligence. Despite the cautions and criticisms of many molecular biologists, behavioral geneticists have begun the arduous process of attempting to correlate variations in the DNA molecule with IQ tests as well as other measures of intelligence. In 1994, a group at the University of London under the leadership of Robert Plomin initiated the IQ Quantitative Trait Loci (QTL) Project. This is an effort to identify common gene variants that are strongly associated with scoring at either the very low or very high end of the standard measures of general intelligence.

Plomin and his colleagues established permanent cell lines from five groups of Caucasian children representing a range from abnormally low IQ (mean of 59) to unusually high IQ (mean of 142). They then performed what are called allelic association studies, asking whether variants in genes were found disproportionately among the five groups. Unfortunately, unless a gene variant is common and strongly associated with performance on IQ tests, this approach requires a brute force effort. To have a reasonable chance of success in identifying variants that have only a small influence on test performance (say, on the order of 1%) scientists must study thousands of DNA markers. Thus far, Plomin has not succeeded in discovering a DNA variant highly associated with IQ, but he is confident that a study of sufficient power (that is, that traces thousands of markers) will do so.

The debate over the proportion of an intelligence measure that is attributable to the effects of genes remains vitriolic. Many psychologists are convinced that analysis of the genetics of intelligence is worthless because it neglects the impact of shared environments. For example, in 1997, researchers in Pittsburgh published evidence suggesting that the much closer correlation in IQ scores between identical twins than between ordinary siblings could be explained largely by shared maternal womb environment. Given that most of brain growth is intrauterine, this is plausible. Perhaps even more interesting is a recent study by scientists at the University of Virginia suggesting that socioeconomic status (e.g., wealth versus poverty) shapes the influence of genes. The study found that in wealthy families genes seemed to explain about 60% of the variance in IQ, whereas non-shared environment explained very little. To the contrary, among poor families, shared environment accounted for about 60% of the variance, whereas gene differences seemed to have little effect!

Although there are few data to support the assertion that particular gene variants account for high performance on standard (albeit controversial) measures of intelligence, hundreds of family studies conducted over the last century suggest that intelligence is highly heritable. The most important weakness of many of those studies is that they often did not adequately control for environmental influences that could have strongly supported success on intelligence tests. This failure has made it socially and politically unwise to suggest that there is a genetic explanation for anything having to do with differences in intelligence across groups.

GENDER

In January 2005, Larry Summers, the president of Harvard University, sparked a firestorm of criticism when he suggested at a closed-door, "off the record" conference convened by the National Bureau of Economic Research, that innate (read genetic) differences between men and women might be partly responsible for the dearth of women professors in mathematics, the hard sciences, and engineering in the United States. One (woman) professor promptly walked out of the conference in protest, and within days many intellectuals were attacking Summers in print for the harm they thought his remarks would cause for young women. The media had a predictable field day, and Summers found himself in the unfamiliar state of cringing before his faculty and apologizing to the world. I have not seen a transcript of his remarks, but by all reports, his suggestion that genetic differences might provide part of the explanation for the dramatic disparity in gender at the high end of human achievement in mathematics and the hard sciences was cautious, constrained, and speculative. Even so, he had to endure months of discomfort and a "no-confidence" vote from his faculty. Summers's social gaffe reminds us that who says something is often much more important than what is said.

Educational psychologists have been aware of and interested in gender differences in mathematical performance for more than five decades. The rapid spread of SAT testing in the 1950s provided a flood of raw data about gender differences to analyze. The weight of evidence provided by hundreds of articles published over the last 50 years clearly demonstrates a gender-based difference in mathematical performance, but it is by no means a simple story. A study of 9,927 intellectually gifted junior high school students in the late 1970s that was published in *Science* concluded that the boys did substantially better than the girls in mathematics. The study design permitted the authors to reject differences in curricula as an explanation, but they concluded that other environmental factors contributed to the disparity.

As far back as 1990, there were more than 100 published studies on gender differences in mathematics, enough to stimulate psychologists at the University of Wisconsin to perform a meta-analysis (an effort to analyze the disparate studies as one unit which might permit new insights derived from the much larger data set). They found that *girls* outperformed boys, albeit by a statistically negligible amount. In subscore analysis, girls outscored boys in computation during the grammar and middle school years. Girls and boys performed equally in word problem solving through the middle school years. However, in high school, the data were much different. Boys began to outscore girls, and the magnitude of the disparity grew with age. Perhaps most relevant to Summers's comments, the more gifted the children in the sample size, the larger the disparity between older girls and boys. Put another way, among really smart kids, boys performed better than did girls in math tests. There is one important caveat to that finding. The size of the gender gap in studies of mathematical talent is substantially less in the studies conducted over the last 25 years than in the 25 years prior to that. This seems to emphasize the importance of environmental factors and to suggest that in a world of true socioeconomic and classroom parity the gender gap in mathematics will be small indeed.

This has already occurred in England. In the mid-1970s, new antidiscrimination laws and a growing body of educational research raised concerns about differential treatment of girls in the classroom. At the time, boys on average did much better than girls on the math portion of the national exam administered to 16-year-olds. The schools implemented programs to address the disparity. A generation later (in 2004), 53% of the girls who took the exam passed, compared with 52% of the boys. In 2003–2004, girls also beat boys on the A-level exams required of college-bound 18-year-olds, with a pass rate of 41% compared to 39%.

The marked improvement in the math scores of English girls dramatically emphasizes the importance of environmental factors in assessing intelligence. However, it falls short of demolishing Summers's provocative speculation. In essence, he suggested that in occupations in which substantial mathematical skill is essential to success (such as math professorships that are almost always held by persons scoring in the top 1% of tested groups), women are and may always be underrepresented. Does the evidence support his contention? There certainly seems to be enough to justify his speculation. In 2000, psychologists at Vanderbilt reported on a 20-year follow-up of 1,975 mathematically gifted young adolescents (aged 12–14) who had scored in the top 1% of the population. In general, the children did exceptionally well in higher education and in their careers. However, career pathways clearly suggested that girls avoided careers which depended on mathematical skills. The gifted girls grew up to be women who chose careers that favored people-oriented disciplines (medicine, teaching), while the boys were more likely to become engineers and mathematicians. The study did not ask whether sex discrimination had been a barrier to entry into the field.

Recent studies have continued to uncover factors that might constrain women from pursuing a career in mathematics. In 1996, a psychologist at Emory University studied a cohort of 66 mathematically gifted middle-school children who were studying algebra in regular classrooms. Although gifted girls outperformed gifted boys, in interviews the gifted girls tended to have less confidence in their abilities than the boys! In 2003, Dirk Dauenheimer, a psychologist at the University of Mannheim in Germany, published a study showing that performance of high school girls on math tests was negatively influenced by "stereotype threat." He showed that in an environment in which girls believed they were not supposed to do as well as boys (math tests), they tended to perform below expectations.

Could there be a genetic basis for the extreme gender disparity at the highest levels of mathematical achievement? It is possible. However, even under the most optimistic scenario, it will be decades before we will be able to resolve the question. First, we will have to eliminate even the thinnest cobwebs of discrimination that restrain young girls as they undergo intellectual development. Next, we will have to extirpate the institutionalized sexism that lingers (albeit at a much reduced level compared to the years when I was in college) in college and graduate school (quite possibly at its worst in departments of mathematics and engineering). Then we will have to wait a couple of decades while young girls grow up in a truly gender-neutral society. Only then (perhaps a half-century hence) might we reasonably conjecture—if in fact it is still the case—whether genetics explains some of the reasons that women do not match men at the highest levels of mathematical achievement.

DISEASES

Dr. Jean-Martin Charcot. (Courtesy of the National Library of Medicine.)

6

Charcot-Marie-Tooth Disease

The small, relatively unknown lake on which I am fortunate to own a lovely old camp was during the first half of the 20th century part of a summer colony that attracted prominent writers, artists, scientists, and more than its share of eccentrics. The father of the poet E.E. Cummings built a house there, and the poet himself lived many summers on a nearby farm (and spent a lot of time in a tree house on the lake's southern shore). In the 1920s, the Cummings family sold the house to a professor at Harvard who was a founding trustee of Skidmore College and a major figure in New York publishing. Among its past inhabitants, the little lake boasts one great scientist—Theobald Smith, America's most famous bacteriologist. In 1893, he proved that Texas cattle fever (babesiosis) was caused by a protozoan organism that is transmitted through the bite of a tick, thus opening up the study of zoonotic diseases. The house he built on the western shore is still in the family. A few miles further west is the home where President Grover Cleveland spent the summers of his later years. According to the woman from whom I bought my place, President Cleveland visited her father on a few occasions. In 1930, Francis Cleveland, the President's son, started what is now the oldest continuously operating summer stock theater in America in nearby Tamworth.

My family and I are newcomers to the lake. Many of my older neighbors live on property that their grandparents purchased. Just a few hundred yards north of my house are the homes of two brothers—one 70, the other 72—whose grandfather owned a good chunk of lakefront long before World War I. The younger, Don (I have changed their names), has made important contributions to providing health care in regions of rural Mexico that could never hope to have a doctor, let alone a clinic. The older, Rob, earned a Ph.D. in mathematics and taught for a time in Ohio. Both have a genetic disorder called Charcot-Marie-Tooth disease.

Depending on their profession, men and women who achieve greatness are remembered in different ways. Presidents of the United States earn a spot on our coins and paper currency. The most sought-after architects leave behind buildings that (at least for the cognoscenti) evoke their memories for many decades. The greatest writers live on in the minds of their readers and are reborn as each succeeding generation discovers them. Hawthorne, Melville, Dickens, and Tolstoy are but a few of the denizens of a literary Valhalla. For a century now, the finest opera singers have been immortalized by recording technology. In medicine, we remember the contributions of the most insightful physicians mainly by eponyms. Their clinical insights live on in the names given to the disorders—or the associated signs and symptoms—that they were the first to describe comprehensively.

Jean Martin Charcot (1825–1893), a French physician who spent many years working at Salpetriere (an ancient asylum that by the 19th century had become a large hospital for the indigent) in Paris, is justly remembered as the founding father of neurology. He was the first to accurately describe amyotrophic lateral sclerosis (still called Charcot's disease in Europe, but known as Lou Gehrig's disease in the United States), he described Charcot's joints, a degenerative foot disease sometimes found in persons with severe diabetes, and his expertise in neuropathology allowed him and a colleague to discover Charcot-Bouchard aneurysm (a small aneurysm in brain vessels that can cause hemorrhage and stroke). At least 15 medical eponyms honor him. Charcot also made major contributions to psychiatry. He was an early student of hypnosis, especially as a possible treatment for hysteria. Sigmund Freud studied with Charcot in 1885, and attributed his interest in hysteria to him. Charcot sometimes conducted his experiments on hypnosis before an adoring lay public that often included members of Parisian high society. His greatness drew many fine young physicians to him, and over the 20 years that he led it, the neurology service at Salpetriere became world famous. One of his most able students was Pierre Marie.

Born the year that Charcot received his first post-graduate position in medicine (1853), Marie, the son of a wealthy Parisian businessman, began his studies under Charcot right after graduating from medical school in 1878. He received an advanced medical degree for an outstanding thesis on a rare neurological disorder called Basedow's disease. Falling perhaps just shy of the greatness achieved by Charcot, Marie too made many contributions in neurology. He made extraordinary contributions to understanding disorders of the spinal column, and in the first decade of the 20th century, he successfully challenged the accepted thinking on the neuroanatomy of speech, showing that some of the areas of brain that were believed to be speech centers had no such function. Unlike Charcot, who died suddenly (probably of congestive heart failure after a heart attack) at the height of fame and happiness, Marie's life was marked by great grief. He lost his beloved daughter to appendicitis, and in his retirement, he lost both his wife and his physician-son to infectious diseases.

In 1886, Charcot and Marie published the first paper describing a familial disorder marked by a slowly progressive wasting of the muscles that begins first in the lower legs and later involves the arms. That same year, a young English physician, Howard Henry Tooth, earned his doctorate at Cambridge University for a dissertation entitled, "The peroneal type of progressive muscular atrophy." His study was a detailed description of families with the same disorder. This was the only major academic contribution that Tooth, who served with distinction as an army surgeon in India, ever made, but it earned him his share of an eponym. Although other physicians had occasionally described patients with what were almost certainly forms of Charcot-Marie-Tooth (CMT) disease, the twin papers from Paris and Cambridge mark the beginning of a research odyssey that is typical of many over the last century, as generation after generation of scientists have tried to understand genetic disorders.

When faced with mysterious conditions, clinical researchers attack the problem by assembling and evaluating the constellation of signs and symptoms that seem to constitute a distinct pathological entity. They also ask whether others have observed similar conditions. Invariably, as years pass and the clinical studies become more refined, the boundary lines defining the entity change. Lumpers become splinters. On closer examination, one disease becomes several, sometimes many, and our understanding of causation, at first little more than a reasonable guess, goes through many iterations. In the case of CMT, it is Tooth who can claim the upper hand on causation. Perhaps because they were already so interested in the role of the spinal column in disease, Charcot and Marie wrongly attributed CMT to a problem in the spine, while Tooth correctly surmised that it involved problems with peripheral nerves (the peroneal nerve supplies the muscles of the lower, lateral side of the leg). Until 1989, a patient organization now called the CMT Association was named the National Foundation for Peroneal Muscular Atrophy.

The tremendous advances in our understanding of CMT disease in the 120 years since its description have occurred over three epochs. Until about 1955, advances in understanding CMT depended on the skill of neurologists who paid ever more careful attention to the clinical course experienced by patients and conducted ever more careful searches for subtle physical manifestations of the disorder. The second epoch, the period 1955–1982, was dominated by electrophysiology. Measurements of the speed with which peripheral nerves conducted signals in muscle, coupled with histological studies of the nerves, provided an important new diagnostic tool. The third and current epoch is dominated by molecular genetics. DNA analysis is the ultimate nosological tool; a single base pair (DNA letter) difference in the involved genes can sometimes be used to differentiate one form of the disorder from another. Over the years, we have learned that the CMT eponym covers a range of disorders with distinct genetic causes.

In the 1890s and in the first half of the 20th century, talented neurologists described many families with what appeared to be CMT, but which included one or more additional features. In 1893, a form of CMT in which patients had unusually thick peripheral nerves was separated out as Djerine-Sottas disease. In the 1920s, a form of the disorder that included tremors was declared a distinct entity called Roussy-Levy syndrome. A major advance occurred in the late 1950s, when CMT was divided into types 1 and 2. To be diagnosed with CMT-1, a patient had to have marked slowing of nerve conduction speed in the peripheral nerves (a phenomenon almost always associated with damage to the myelin sheath that wraps and insulates the nerves). Patients with CMT-2 have relatively normal conduction speeds, but a marked decrease in amplitude of the signal, a phenomenon associated with degeneration of the axon (nerve body) rather than the insulation that wraps it. By the 1970s, especially in Europe, experts had adopted the generic term, Hereditary Motor and Sensory Neuropathies (HMSN), to describe at least seven distinct clinical forms of CMT.

The great advances brought about by molecular genetics have been in understanding causation. The latest edition of the leading textbook in medical genetics (which compromises by giving the chapter the title, "Hereditary Motor and Sensory Neuropathies," but retains the fundamental distinction between CMT-1 [myelin damage] and CMT-2 [axon damage]), identifies 18 distinct genetic subtypes of the disease first described by

Charcot, Marie, and Tooth. All are single-gene disorders, obeying one of the three (dominant, recessive, X-linked) forms of inheritance. Eleven are associated with slow nerve conduction and myelin damage, and seven with axonal loss.

Although CMT-1 and CMT-2 can now legitimately be considered as two nosological baskets containing at least 18 disorders inherited in three different patterns, the majority of patients have an autosomal dominant form of the disorder, and almost 80% of them are affected because they have a duplication of a gene called PMP22 (peripheral myelin protein 22), which resides on the short arm of chromosome 17. A duplication arises due to a copying error during the formation of sperm or egg. Until the advent of the molecular era, human geneticists did not include duplication errors as a cause of dominantly inherited disorders, but now we know they are not all that uncommon. On sheer statistical grounds, it is likely that this was the genetic disorder that affected the families that the three 19th-century neurologists first described.

Remarkably, the age of onset, severity, and clinical course of CMT caused by a duplication of the PMP22 gene can vary widely even within the same family. Although one can diagnose CMT by nerve conduction studies (and of course with DNA testing) in early childhood, symptoms usually do not appear before age 6 and can arrive as much as a decade later. On the other hand, it is rare for the disease to develop in a person born with a one-in-two risk (e.g., with an affected parent) after age 30.

When diseases are uncommon, patients are the best teachers. Although I was familiar with the textbook descriptions of CMT disease and have occasionally seen patients in a hospital clinic, I knew that I could draw a clearer picture of the disorder if I interviewed Don and Rob. Even though they have a less common, X-linked form of the disorder (in which women who carry the causative mutation on one of their two X chromosomes are not affected, but have a one-in-two chance that a son will be affected), the natural history is not much different from that of the more common dominant forms. The most surprising aspect of interviewing these two interesting men is how differently the same disorder affected their lives.

My tale really begins with their mother. A brilliant woman who earned a degree from Harvard when few women went to college, she was well aware that a muscle-wasting disease that affected only men ran in her family. In her teenage years, she vowed not to marry so as to avoid bearing affected sons. For a time it looked like she would keep her vow, but when she was well into her 30s, she fell in love with and married a lawyer from Ohio. Hoping for girls or unaffected sons, Mary twice lost the genetic game of chance.

Her first-born son, Rob, was diagnosed as having the disorder at about age 6 when he developed the waddling gait typical of affected children who compensate for weakness in their lower legs by using their hip muscles to pull themselves along. At the time, Don, two years younger, had not yet manifested any signs of the disorder. As he recalls it, his mother, grieving over the disease in her firstborn, would "not allow" it to affect her second son. In his memory, her denial was overwhelming.

Her reaction to the disease set her sons on two very different life paths. From the outset (according to Don) she forced Rob into the sick role, while Don was ordered to be healthy. Don did not learn he had CMT until age 9, about three years after he developed symptoms. He made the discovery one day when he was sneaking a look in his mother's papers and found a three-year-old letter from the pediatrician who had diagnosed CMT in Rob, informing her that he thought Don was also affected. More than 60 years later, Don recalls that it was at about age 6 that his mother started taking him (but not Rob) to a new pediatrician who knew nothing about CMT. The reason was obvious: She did not want Don labeled with the disease.

Don is convinced that being affected with CMT shaped the course of his life. His memories of childhood are replete with images of taunting. He recalls that at age 8, as he was losing strength in his lower legs, schoolmates walked behind him in single file, calling "Quack, quack, quack," mocking his duck-like gait. At summer camp in Maine when he was 10, he struggled to master swimming. At the camp's closing ceremony, he was sure he would get the prize for "most improved." Instead, he got a prize for being the "clumsiest." That "really hurt," he recalls. In junior high, some of his classmates took to calling him "rickets" (the common name for a vitamin D deficiency disorder that causes bowed legs and, at that time, a common cause of an awkward gait), and some teachers even used the nickname.

One of the most important lessons that Don took from his childhood struggle with CMT was to question authority. Starting at about age 9, when his mother could no longer deny his problems, Don was repeatedly fitted with braces to assist him in walking. The problem, of course, was that in the 1940s, few physicians knew anything about how to help children with CMT. The braces were harmful. They increased the frequency with which he twisted his ankles and some cut into his flesh. He recalls those years as a long battle with his mother, who kept insisting that the doctors knew more than he did. Sadly, she was wrong.

There were other, more poignant events. Don's father came from a family of musicians. At age 6, Don took up piano and advanced rapidly. By age 8, however, he was losing his playing skills. The wasting muscles in his fingers would no longer follow his commands. Turning to pastimes that did not require so much fine dexterity, Don took up tennis, but by high school, he lacked the forearm strength to return the ball with speed and finesse. As friends no longer found him a worthy competitor, he gave up the game. By high school, Don, whose stick-like lower legs and crooked hands made him self-conscious, had became a loner. This, he thinks, is what led him to develop a lifelong fascination with nature, a solitude into which he could retreat.

Unlike his older brother, Rob, who from the first was forced to take a passive approach to his disorder, Don fought it. He became obsessed with physical fitness, sure that regular, strenuous exercise would delay loss of function. When he was in his mid-20s, an age by which Rob was having real trouble getting around on his own, Don vowed to ride a bike from Boston to India (where he laughingly recalls he thought he would find people who would ignore his failing muscles and appreciate his mind). Although he nearly died along the way from dehydration caused by dysentery, he made it to India. He did not find the higher level of thinking he sought. Shortly after this amazing journey, Don, who had taken a degree in biology, decided to spend his life helping people to overcome disabilities. For several decades now, he has worked in villages in rural Mexico, helping to teach common folk to build wheelchairs, make prostheses, and obtain education. Because of connections he had at hospitals in California, he has obtained free orthopedic surgery for more than 300 children! Don has wonderful stories about disabled kids he has helped who have succeeded far beyond anyone's wildest dreams. One boy with polio whom he met when he was 3 and who had never walked because the village had no crutches, is now a third-year medical student. His motto in life, one that he uses in his work with people with disabilities, is "Look at my strengths, not my weaknesses."

Don turned 70 the week before I interviewed him. To celebrate, he climbed Mount Chocorua, a 3800-foot peak in the Sandwich Range, for the ninth time that summer! How could that be? Don would be the first to say that it was not easy. He has had five major surgical procedures to try to restore or protect function in his hands and feet. About 20 years ago, he had three operations to transplant several tendons from his feet to his hands in an effort to counter the severe contractures of his fingers. Today, both hands function fairly well, but the left is much better than the right. He describes the surgery on his left hand as "highly successful," and attributes the different outcomes to the superior work of one surgeon over the other. One glimpse at his hands is confirmatory. Although both hands are small and flat, three of the fingers on the right are so contracted that he cannot use them. None of the fingers on his left hand are contracted. Three years ago, he had surgery to fuse his lower legs to his ankles (a procedure that counteracts foot drop). But for that surgery, Don says, he might not be able to walk without crutches, let alone climb Mount Chocorua.

What does a 70-year-old man with Charcot-Marie-Tooth disease look like? Don is about 5 feet, 7 inches tall. He is bearded and slender. He walks with a slight stoop and often uses a walking stick. Don uses a brace made for him by a young Mexican man to assist him in walking and to counter his great risk for twisting his ankle. The only deficit that is immediately obvious is the claw-like nature of his right hand. He holds a glass of lemonade with his palm rather than his fingers. If Don is wearing a short-sleeved shirt, even an unpracticed eye will note his thin, flat forearms with muscles bulging at the elbow.

The evening before I wrote these words, Don, Christopher, my 11-year-old son, and I swam about 400 yards out to an island and back. When he is in his swimsuit one immediately notices that Don has (in his words) "nothing below my knees." His lower legs are stick-like and his feet turn inward. Don swims with a modified backstroke that relies heavily on his shoulder strength. His upper arms and torso are muscular, even robust. Don can do far more chin-ups than can I. No one knows why CMT spares the proximal muscles.

The contrast with Rob is dramatic. At 72, Rob is wheelchair-bound and has an indwelling bladder catheter (for reasons not directly related to the effects of CMT). It is unusual for persons with CMT to lose the ability to walk. A few years back, Rob, who then could still walk with crutches, was struck by a car and lost a leg to amputation. His remaining leg is not strong enough to support efforts to use crutches, or at least Rob has refused to attempt to do so. An extremely talkative fellow, Rob often grips one's arm as he tells his stories. His hands are claw-like with the fingers flexed into the palm, so his grip is with his knuckles. The backs of his hands are flat with little valleys between the bones. His forearms seem to have no flesh at all until just before the elbows, when they bulge out suddenly. Unlike Don, Rob requires almost constant assistance. For example, he cannot cut the food on his plate.

If there is a theme to Rob's history, it is the "ignorance" of the medical profession. After hearing his story, one can sympathize with his perspective. Sometime in the late 1930s, Rob's parents took him to Boston for an evaluation by a prominent Boston pediatrician. As Rob remembers it, after three days of tests, the pediatrician diagnosed Rob as having Friedrich's ataxia, a neuromuscular disorder that, because of its effect on heart and respiratory muscles, was often fatal by age 20. Although the doctor apparently said CMT was also a possibility, Rob claims to have spent his childhood thinking he would not live past 20. This may account for some of his daredevil behavior. He reports that he carried a switchblade knife for protection against bullies, and that he once even used it to fend off taunting schoolboys. At age 18, even though he needed special shoes so he could operate the plane, Rob flew solo over Conway, New Hampshire in a J-3 Piper Cub. Although the army would never have taken him, he seriously considered enlisting to fight in the Korean War.

Rob is a brilliant, but eccentric, individual who showed very early promise. He claims to have developed an interest in electricity at the age of 3. After a strong showing as an undergraduate, he was accepted into a Ph.D. program at Princeton to study with John Wheeler, one of the greatest physicists of the 20th century (and at this writing still alive and approaching the century mark). But Rob left after only two years, in part because he did not stay focused on his assigned projects. He eventually completed a Ph.D. and taught for a while at the University of Cincinnati. But his academic career fell short of his early promise. It may be that there is no connection between his struggle with CMT and his career, but Rob does seem to be obsessed with his disorder. He told me that he is sure that the reason it is not worse is that, when he was about 12, a physician experimented on him by giving him daily injections of histamine (a drug which makes you flush red and can cause a monstrous headache). I can find no evidence at all that such an approach has ever been attempted by anybody else.

What does the future hold for people with CMT? Armed with the tools for molecular dissection, geneticists and neurologists are doing genotype–phenotype association studies, work which correlates particular mutations within the genes that cause CMT with the clinical picture of the patient. This is tedious, difficult work, but it sets the stage for the end of the third epoch. That will be reached when scientists can explain how it is that a particular mutation in a specific gene leads to a defective protein which compromises certain peripheral nerves in a manner that gives rise to the clinical picture we have called CMT for more than a century. We are beginning to get some answers. For example, we now know that persons with duplications of PMP22 have a defect in a part of the nerve axon called the cytoskeleton that slowly causes nerve degeneration and loss of fibers. Other work has shown that a particular mutation in PMP22 also causes deafness. In 2002, a rare form of CMT was described in which affected persons also develop paralysis of the vocal cords even though other muscles in the neck and head are unaffected!

One totally unexpected recent finding is that at least some people with CMT are at great risk if they develop cancer and are treated with a drug called vincristine. Several relatively young people under treatment for cancer have nearly died when the drug caused a profound failure of their musculature, including their respiratory muscles. After the fact, a careful family history and close examination of the feet and lower legs uncovered early clinical signs of CMT. No one knows why persons with CMT are at special risk from vincristine, but it now should be a standard of care to evaluate patients who are candidates for treatment with it to rule out CMT.

It is hoped that the fourth epoch in the history of CMT will soon open. That will be the era in which treatments for this still untreatable group of diseases emerge from the research laboratories. As they come to understand the causative genes, scientists will be able to find new targets for drug development. For example, in the fall of 2004, researchers at the University of Edinburgh showed that one form of CMT is caused by mutations in a gene that makes a protein called periaxin. They proved that the periaxin protein is important in maintaining the insulating material called myelin that ensheathes nerve axons. In people who have CMT due to mutations in the periaxin gene, the myelin becomes discontinuous and nerve impulses slow. The periaxin protein could point the way to a new drug that slows the progress of this form of the disease.

This process will be arduous. Because only about 1 in 2,500 persons is affected, there is no incentive for the large pharmaceutical companies to invest in drug research for CMT. The more likely scenario is that early findings in academic laboratories will lead to the creation of one or more biotech companies deeply focused on developing a treatment. Such efforts depend heavily on venture capital, and the ability to raise the necessary tens of millions of dollars depends greatly on the investors' appetite for risk in a difficult market. Many companies have been built around a single pharmaceutical product, and many of them have failed. But there have been a few spectacular successes, a fact that keeps the financing opportunities alive.

The family of disorders known as CMT disease will be especially challenging to cure. However, because they are slowly progressive disorders caused by defects in a single protein, it is plausible that over the next two decades scientists will discover how to significantly limit the pace with which people lose nerve function. Oddly, there are very few eponyms in medicine that describe cures.

Nancy Wexler, Ph.D., played a key role in finding the gene that causes Huntington's disease. (Circa 1986; Courtesy of Cold Spring Harbor Laboratory Archives.)

Huntington's Disease

Where do genetic diseases come from? Each begins with a change in a DNA sequence in a germ cell (egg or sperm) that unites with its counterpart to create an embryo. Usually, the change occurs in a part of a gene that codes for a protein (or, less commonly, a nearby DNA segment called the promoter that influences the production of the protein), impairing it in some way. Many agents—including ionizing radiation, chemicals, and errors by the DNA replication enzymes—can cause mutations. Overall, however, the mutation rate is quite low. Sometimes, one DNA letter is substituted for another. This changes the message that is used by the cell as the blueprint for protein production. The result may be an incompletely formed (truncated) protein that is almost always a serious problem. It could also result in a change in just one amino acid in the sequence that makes up a protein, an alteration that can be benign or devastating, depending on the impact it has on the three-dimensional structure of that protein. Many other kinds of changes, including duplications, deletions, and transpositions of the genetic material, can also result in genetic disorders.

Of course, for a genetic mutation to give rise to a disease, the germ cell in which it resides must unite with another and form a human embryo that is capable of survival, at least for a time. There are probably many genetic abnormalities of which we are unaware simply because they are incompatible with fetal life. These invisible disorders probably cause many early, spontaneous miscarriages. Indeed, it is possible that many embryos die for genetic reasons before women even realize they are pregnant. For a disease to be recognized, those burdened with it must come to the attention of physicians.

Most genetic diseases that are due to mutations in a single gene are transmitted in one of three ways (taught in school as the laws first clearly described by Gregor Mendel, the Czech monk who studied the inheritance of traits in pea plants). Recessive disorders are those that only manifest if an individual inherits *two* mutated forms of the same gene, one from each parent. Usually, the diagnosis of such a disorder is a complete surprise to the family. Because mutations are uncommon, it is possible for a recessive gene to travel through many generations before a germ cell from a family member who carries it unites with a germ cell carrying a comparable mutation in the same gene.

Dominantly inherited genetic disorders and X-linked recessive genetic disorders are not so subtle. Assuming the condition lets them live to reproductive age, persons with a dominant disease have a one-in-two risk of passing the deleterious gene to each of their offspring (regardless of whom they marry). In general, dominant disorders tend to be less devastating and to manifest at an older age than do recessive disorders. Over the generations, family members become intimately familiar with the nuances of these disorders. The mutations that cause X-linked genetic disorders have a still different impact. They essentially spare females (in whom the X chromosome without the mutation counters the impact of the X chromosome with the mutation), but affect one-half of the sons who are born to the carrier females (a boy has an equal chance of having inherited either the normal or the mutated X chromosome).

In 1893, William Osler, the chief of medicine at Johns Hopkins and perhaps the most legendary physician in the history of the United States, wrote a paper in which he described the course taken by a hereditary form of chorea in two families. The term chorea connotes the inability to control the action of the muscles. Osler noted that in individuals in whom this progressive disorder was well advanced, the arms, legs, and even the facial muscles moved almost unceasingly in a bizarre, random fashion. Osler was among the very first physicians to perform an autopsy on a person with hereditary chorea and, as far as I can determine, was the first to link specific brain abnormalities with the disorder. He observed that parts of the brain, especially the caudate and the putamen (which lie near each other in an area called the basal ganglia at the bottom of the brain) were atrophied, suggesting the death of certain kinds of cells.

Given his contribution, why do we not call this Osler's disease? In his 1893 paper, Osler acknowledged that in 1872 a young Ohio physician, George Huntington, had written about a large group of people with what appeared to be the same disorder, all of whom lived at the eastern end of Long Island. Dr. Huntington's interest in the disease had begun in boyhood when he traveled with his physician father to visit patients. Years later, as a young doctor, he accompanied his grandfather and father, both of whom practiced medicine on Long Island, to examine these folks. Although the younger Huntington wrote just four long paragraphs about these patients (in an addendum to a paper on nonhereditary forms of chorea), the noble Osler credited him with the description that earned the eponym. In 1908, recalling the report published by George Huntington, Dr. Osler opined that, "In the history of medicine there are few instances in which a disease has been more accurately, more graphically or more briefly described." Russell DeJong, a diligent medical historian, has proven that at least five physicians described patients who almost certainly had HD before Huntington's paper was published by *The Medical and Surgical Reporter of Philadelphia* on April 13, 1872. This, however, does not challenge the award of the eponym. Their clinical descriptions were less complete and, because they went largely unnoticed, they did not advance clinical practice.

Huntington recognized that the disease was hereditary, but he failed to grasp the notion of a dominant disorder, thus just missing scientific immortality. His 1872 paper appeared 28 years before the rediscovery of Mendel's work delineating the laws of inheritance. Huntington (who called the patients "shakers") wrote that if a parent was affected, "one or more of the offspring invariably suffer the disease." This is incorrect. If an affected parent has two children, there is a 25% chance that neither will be affected. On the other hand, Huntington accurately described the psychiatric complications that almost always accompany the physical manifestations of HD.

Osler, whose work also predated the rediscovery of Mendel's work, was primarily concerned with delivering a full clinical picture of the disorder and correlating it with changes in the brain, so he did not compile sufficient evidence to prove an inheritance pattern. Overwhelming evidence of dominant inheritance came in 1916 when Charles Davenport, who was greatly influenced by the rediscovery of Mendel's work in 1900 and became one of the nation's first human geneticists, published an extensive study of how the disorder, now called Huntington's chorea, traveled in families on Long Island. Among 962 patients, he found only 5 who were not clearly the children of an affected parent.

In 1932, a scientist named Vessie, with the cooperation of some of the descendants of the people whom the Huntingtons and Davenport had examined, constructed an elaborate genealogy. He was able to show that nearly all HD patients on the east coast of the United States, more than 1000 persons, were descendants of a few individuals who had lived in a tiny East Anglican village called Bures St. Mary in Suffolk, England in the 16th century and who had immigrated to the colonies in 1630. Huntington's disease may well have been the first dominantly inherited disorder carried by Europeans to North America. Vessie showed that in one branch of this family the disease had appeared in 12 consecutive generations. Since Vessie's work, other researchers have traced the flow of the HD mutation across the planet. Because some members of the original family were seafaring men, it is hardly surprising that the disease popped up far from England. For example, there is a concentration of at-risk families in New South Wales.

For a long time, geneticists thought that all living affected persons were descendants of a Suffolk man, in effect surmising that the mutation had only "taken hold" once in human history. More recent research (to which I shall return) suggests that in some patients (perhaps 1–5% of the total) the disease arises de novo. Thus, the 5 of Davenport's 962 patients without a family history could represent new mutations.

Today, after more than a century of study, neurologists are painfully aware of the clinical details of this dreaded disorder. The age of onset varies; symptoms usually appear in the late 40s and 50s, but the range is wide. Huntington's disease has virtually complete penetrance, meaning that if a person who has inherited the mutation lives long enough, he or she will develop the disease. One child has been diagnosed at age 2, and at least one person known to have the causative mutation had not yet manifested symptoms at the age of 90. Once symptoms become manifest, this degenerative brain disorder runs an inexorable downhill course that usually leads to death in about 15 years. In the early ("transitional") phase, patients often notice a general restlessness and a loss of fine motor skills such as those used in sewing. They also may note slight slurring in their speech. As the disease progresses, the patient develops choreic (involuntary) movements that are nearly constant while he or she is awake. The face is a sea of pouts, twitches, grimaces, and frowns. Affected persons frequently fall, eventually lose the ability to walk, and, ultimately, even to feed or dress themselves. In advanced stages, the speech may be so slurred as to be incomprehensible. Many even often lose the ability to swallow their food. Impairment of memory occurs fairly early. Irritability, severe depression, and even delusional thinking are not uncommon. The average age at death is about 60 but varies widely. Patients often die of pneumonia or other infections to which, in their wasted state, they are vulnerable. Affected individuals can feel so helpless that it borders on terror. Historically, patients in the advanced state of HD have had a suicide rate several times higher than the average for their age group.

I first became acquainted with HD about 30 years ago when I met two extraordinary people, Marjorie Guthrie and Nancy Wexler. Marjorie was the widow of the great folksinger, Woody Guthrie, who died of HD. Nancy, at the time, was a young clinical psychologist whose mother was afflicted and who was at one-in-two risk for having inherited the HD mutation. In those days, Marjorie, an indomitable woman, was engaged in a one-person lobbying effort with the U.S. Congress to secure more research funds for Huntington's disease. In 1977, she succeeded. As far as I know, she is the only person who single-handedly convinced the U.S. Congress to pass a law (that even included an appropriation). It was an act to allocate federal funds to support research into the cause of HD. Even in the late 1970s, Nancy Wexler was extremely concerned about the impact that our nascent ability to diagnose genetic disease would have on society. She was among the first to worry about the impact of predictive knowledge on the emotional health of an individual and one of the first to worry publicly about what would come to be called (by about 1986) genetic discrimination.

While Marjorie busied herself with issues in research funding and public education, Nancy became ever more involved with research on understanding the molecular basis for the disease. By 1978, rapid advances in the ability to study DNA made it theoretically possible to find the gene associated with the disease, something that would have been considered a fantasy a decade or so earlier. For dominantly inherited disorders, the initial approach is to try to find a stretch of DNA that, within a family, seems to be co-inherited with (track with) the disease-causing mutation. For a small stretch of DNA to be a helpful marker, it must almost always be present in persons with the disease allele, but almost never present in unaffected at-risk relatives (the rule is not absolute because of a phenomenon called crossing-over in which gene segments are exchanged during formation of germ cells). This can only occur when the DNA probe is located sufficiently close to the HD gene that the normal shuffling of DNA which occurs during formation of egg and sperm does not separate the two. If such a probe could be found, then one would have a rudimentary genetic test to predict in at-risk families the likelihood that a particular individual would develop the disease. In the early going, this was thought to be at least as difficult as finding a needle in a haystack. A complete set of human DNA (in the sperm or egg) is about 3 billion chemical letters long and contains about 25,000 genes!

Perhaps the greatest concentration of persons affected with, or at risk for, HD lives in a collection of villages along Lake Maracaibo in Venezuela. In effect, the population is an extended kindred that numbers over 14,000. Beginning in the late 1970s, Wexler, along with colleagues in molecular and population genetics, set out to find the needle in the haystack, to find the gene that caused the disease. At the time, the effort seemed quixotic. She and her colleagues had to convince this largely impoverished and un-educated people, few of whom had ever received good health care, to join a research enterprise that offered them no promise of cure. They then had to construct human pedigrees of unbelievable complexity involving thousands of relatives. They also had to perform neurological examinations on at-risk persons and obtain blood for DNA analysis. Back in Boston at the Massachusetts General Hospital, a young molecular biologist named James Gusella and his team would perform the DNA studies and statistical analysis to see whether any of the various DNA probes they had assembled would track with inheritance of disease.

To everyone's surprise, the team quickly found the needle. The eighth DNA probe that they tried tracked with the disease mutation, and they were able to show it was located on the short arm of chromosome 4. Their report that appeared in the prestigious journal *Nature* in 1983 is a landmark in modern genetics. Gusella was soon being called "Lucky Jim." Since he had no rationale with which to focus his search, he had simply won the roll of the dice. On statistical grounds, he might easily have studied hundreds of probes without finding the sought-for correlation. His good fortune takes nothing away from the extraordinarily hard work and dedication that has characterized his career.

Although crudely mapping the HD gene was a big job, it paled before the challenge of actually finding the gene and isolating (cloning) it for study. It took a large, collaborative team of scientists another decade to accomplish that, and when they did, the gene was full of surprises. By comparing the sequence of the gene in normal individuals to the sequence in persons with HD, the scientists found that Huntington's disease is one of a newly described class of genetic diseases called "triplet repeat" disorders.

In the DNA language, three nucleotides code for one amino acid. DNA has four nucleotides (chemical letters often called A, T, G, and C) that can be arranged in a total of 64 (4 × 4 × 4) three-letter combinations. But the DNA code is redundant; three or four different three-letter combinations code for the same amino acid. Also, three of the three-letter groups constitute "stop" codons; they code for a message to terminate construction of a protein chain.

A triplet repeat disorder is one in which a three-letter string of DNA somehow gets amplified during the formation of egg or sperm. Today we know that people who carry the normal version of the HD gene have between 10 and 35 copies of the triplet called CAG (which codes for the amino acid, glutamine). For some reason, people with 36 or more copies of the CAG in that gene develop HD. Some affected individuals have more than 100 copies of the CAG triplet in their DNA. Thus, at the molecular level, HD is caused by an expansion of CAG in the gene that in turn creates an unusually long string of glutamines in a protein we now call huntingtin. Another surprise is that, generally speaking, the larger the number of repeats, the earlier the age of onset of the disease. Yet, the number of repeats does not generally correlate with the severity of the disorder. Even more surprising is that in triplet repeat disorders like HD, the gene appears to have undergone a "gain of function" rather than the loss of function that one would anticipate.

The discovery that HD is a triplet repeat disorder has helped to resolve the debate over whether apparently sporadic cases of HD (in persons with no known affected relatives) are really the result of new mutations or could be explained in other ways (such as non-paternity). In a careful study of 16 different cases of "sporadic HD," it has been shown that in each case the expanded triplet repeat came from the normal father of the affected individual. Another study found that although it was below 35 triplets in length, the repeating unit in the unaffected parent tended to be much larger than in most people (on average 29 as opposed to 14). This has led to the theory that normal men with alleles in the low 30s are at risk that this triplet repeat will undergo further expansion during formation of sperm cells. This is probably what happened to cause HD in Davenport's five patients who lacked a family history. On the other hand, we now also know that it is possible to be born with a few more than 35 repeats and not develop the disease.

The discovery of the G8 probe made it possible to offer predictive testing to persons at risk. Until the causative gene was cloned in 1993, the test could substantially, but not absolutely, restate the odds. Rather than living with a 50% risk, individuals could learn with more than 90% likelihood whether or not they had inherited the disease allele. Among clinicians and at-risk persons, the possibility of testing caused consternation. Would a positive (higher risk) result provoke depression or even suicide? Would a negative (lower risk) result offer false reassurance or create survivor guilt in those who learned a beloved sibling had not been so lucky? Should physicians require psychiatric evaluation of at-risk persons before providing them with the test? Should the test be used in prenatal diagnosis? Should a test be offered to one person if the result would render diagnosis on another (for example, one of a pair of identical twins or an at-risk adult whose at-risk parent has not been tested)? What would the impact of a positive test result be on the spouse of an at-risk individual?

For someone who has never seen a person suffering with this terrible disorder, it is nearly impossible to imagine its physical and emotional burdens. In HD families, individuals who are born at one-in-two risk (those with an affected parent) from an early age view life through a prism of risk unlike that known to the rest of us. Even the most optimistic among them cannot envision a healthy, robust old age. Several decades ago, I had dinner with a young woman whose mother had HD. She told me that whenever she was doing anything that remotely resembled a date she always promptly told men that she was at one-in-two risk for HD, and that if she married she never would have children.

I well recall meeting a man at an HD conference whose father had died of the disorder. This handsome, charming fellow in his late 30s gave a talk about living at risk. Some persons who are at one-in-two risk are able to contain or deny that fact and live normal lives. Most cannot do this. This man had, he told us, from the age of 16 lived as though he was going to die of HD by the age of 40. He was an inattentive student, quick to argue with his bosses, and uncommitted in his relationships. He had a vasectomy when he was 21. He quit jobs on a whim. But at the age of 30, he fell in love, a snare that he had sworn to avoid. He told the woman all about HD, but could not drive her away. He married, and his wife, who wanted children, eventually convinced him to take a genetic test that had been developed to determine pre-symptomatically if a person at risk had won or lost the genetic toss of the mutational coin. In contradiction to his most deeply held belief—a conviction that had guided his conduct in life, he learned that he had not inherited the disease mutation. He had won the coin toss. The good news certainly changed his life. As he put it, "I had to reverse the vasectomy, be more polite at work, pay off my debts, and start saving for my kids' college tuition." The audience roared with laughter.

For several years, a coterie of geneticists who directed the few laboratories which offered predictive testing for HD adhered to a relatively rigorous screening process, essentially refusing to offer the test unless an at-risk individual first underwent psychiatric evaluation. In addition, with rare exceptions they refused to test persons under the age of 18 or to offer prenatal diagnosis. During the late 1980s and early 1990s, research on persons at risk for HD who decided to undergo "linkage-based" (the statistically based test that was used before the gene was cloned) predictive testing showed that the half who received good news benefited substantially, while very few of the half who were told that they would almost certainly develop HD suffered serious psychological problems. A number of surveys show that few people (even those who test positive) regret having been tested. By 2003, there seemed to be rough consensus among involved clinicians that predictive testing posed little danger of serious psychiatric events. However, in 2004, a group from Johns Hopkins published data on a one-year follow-up of more than 150 persons that found depression was about twice as likely among persons who had tested positive than among those who had tested negative.

The most intriguing fact about the history of predictive testing for HD is how few of those who are at risk decide to be tested. A comprehensive retrospective study of HD testing in Canada from 1987 through 2000 found that only 18% of at-risk adults took the test. The number became slightly higher in the late 1990s after the gene was cloned and CAG repeat testing became available. Given that the disease is untreatable and the age of onset is uncertain, most at-risk persons prefer not to know if they were born with the mutation. Furthermore, very few people who know they have the mutation and who decide to have families choose the option of prenatal diagnosis and selective abortion. As of 2004, I know of only a handful of prenatal tests (that identified 12 affected fetuses, of which 11 were aborted). In having children at one-in-two risk, affected parents express the hope that 30 years hence there will be either a good treatment or a cure for the disorder. The Canadian study also investigated the use of HD testing as a diagnostic tool in persons with questionable clinical symptoms but without a family history. Their findings confirmed the suspicion that up to 10% of HD is due to new mutations, much higher than had been recognized.

The courage and optimism shown by HD families today contrasts sharply with the far more pessimistic outlook a few decades back. This is in no small part due to Marjorie Guthrie, who led the battle to remove the eugenic taint from this disorder. Sadly, Davenport, the man who established that HD was a dominant disorder, was also a leader of the eugenics movement in the United States during the first quarter of the last century. This movement, which included prominent financiers, politicians, physicians, attorneys, and social critics among its leadership, advocated that some individuals (primarily the mentally retarded) should not reproduce. For a time, eugenics groups had great success in shaping social legislation, including compulsory sterilization laws in more than a dozen states and restrictive federal immigration policies. Davenport was their most influential scientist. He frequently pointed out that HD was an excellent example of a genetic disorder that could be wiped out in one generation. If at-risk people chose adoption, instead of having offspring of their own, the disease would nearly vanish. The idea never took hold in the United States (where the focus was largely on sterilizing institutionalized retarded persons), but it did in Nazi Germany. Risk for HD was among the conditions for which the German "racial hygiene" courts could order sterilization.

With the cloning of the HD gene in 1993, hope for advances in understanding the disease at the molecular level and for developing a treatment soared. But the going has been tough. The last decade has not yielded a single meaningful therapy, and scientists are still struggling to learn what the relevant protein (huntingtin) does in normal cells. However, they have learned that the hallmark of HD is the death of a type of brain cell called the medium-sized spiny neuron in an area called the striatum. They have created rat and mouse models of HD, and they have preliminary evidence that certain drug interventions modulate the disease in animals. Such advances continue to fuel the hopes of the 30,000 HD patients in the United States, as well as those of the 150,000 who are at one-in-two risk. But with hope has come impatience. This has generated some unconventional research.

In 2003, Lavonne Veatch Goodman, whose first husband died of HD, and her second husband, Nathan Goodman, a prominent computer scientist who works in bioinformatics, teamed up to start a not-for-profit called Huntington's Disease Drug Works (HDDW) with the highly focused goal of greatly accelerating novel therapeutic studies in persons with the disease. Eschewing placebo-controlled trials and other standard approaches to drug research, the Goodmans are moving rapidly to enroll HD patients in clinical trials of one drug (cysteamine) and five natural supplements (blueberry extract, trehalose, omega-3 fatty acids, creatine, and coenzyme Q10) to see whether they can find evidence of short-term benefit. The CoQ10 experiment is especially intriguing because the supplement has been shown to improve motor activity in transgenic mice that develop HD-like symptoms after having a version of the human HD gene engineered into them in embryonic life. Due to its highly unconventional approach to clinical trials, HDDW has elicited criticism within academic medicine. However, given the small number of persons affected with HD, a search for a new treatment is very unlikely to provoke much interest in the pharmaceutical industry. An HDDW success could, however, greatly interest a small biotech company to enter into an agreement to move the project forward. It is the plight of patients with rare disorders like HD that stimulated Congress to approve the "orphan drug" law which provides strong financial incentives to companies that choose to work in disease areas with only a small number of patients.

The most promising area of therapeutic research at the moment in animal models of HD is the transplant of healthy brain tissue from fetuses to the brains of adults with the disease. Experiments in rats with experimentally induced HD have shown that fetal transplants sometimes have a protective effect. HD is one of several diseases in which the most promising therapeutic pathway in humans is hindered by the ethical and legal debates over the permitted uses of stem cells. If George Huntington could visit with us today, he would, I am sure, be delighted by how much we have learned about the mysterious disorder he described. He would also be surprised by how much influence this rare disease (affecting about 1 in 10,000 persons) is having on the manner in which researchers think about how to develop treatments for intractable disease. But he would be horrified to learn that controversy over use of embryonic stem cells was delaying potentially lifesaving research efforts.

Students at Gallaudet College using sign language. (Courtesy of John Consoli/Gallaudet University.)

Deafness

A child's joyful laugh, a loon's mournful call, a Mozart piano concerto, the crack of David Ortiz's bat against a Yankee fast ball, and the countless other sounds that help us understand, navigate, and take pleasure in the world begin as pressure waves that flow through the external auditory canal and strike the tympanic membrane (eardrum) that marks the boundary between the outer and the middle ear. The resulting vibrations are transmitted through the three tiny bones (hammer, anvil, and stirrup) in the middle ear. It is in the inner ear that sound vibrations are transformed into thousands of electrical signals which are relayed through the auditory nerve from which they radiate along neuronal pathways, reaching in a few milliseconds a region in the temporal lobe of the brain called the auditory cortex. There, newly integrated information quickly generates signals that travel to other parts of the brain where they are filtered by memory and wrapped in emotion, evoking joy, calm, fear, or elation.

Virtually all our knowledge of any aspect of the human body begins with its anatomy. Given the location of the middle and inner portions of the ear, it is not surprising that it was among the last anatomical niches to be fully explored. To explore the middle (and, subsequently, the inner) ear, a first-year medical student must patiently drill a tiny shaft through the petrous bone, the rock-like bulge that you touch if you place a finger just behind the outer ear. I recall vividly the snowy afternoon nearly 30 years ago when I (a generally clumsy anatomist) felt a spark of triumph upon reaching the middle ear of my cadaver without having damaged the three little bones that for seven decades had carried sound waves from the outer world to his inner ear. Anatomists first described these two tiny cavities (the volume of the middle ear is about one-half of a cubic inch) four centuries ago, but scientists lacked even a rudimentary understanding of how the structures worked until late in the 19th century. It was during the 1870s that Richard Heschl, an Austrian pathologist, elucidated that a small prominence in the temporal lobe of the brain (today known as Heschl's gyrus) acted as a central processing area to correlate sound with meaning.

The key structure in the inner ear is the cochlea, a snail-shaped organ that is responsible for both our sense of balance and our ability to hear. We maintain our balance in large part because of three tiny (1/20th inch in diameter) semicircular canals in the vestibular portion of the cochlea. They contain a fluid called endolymph, the ebb and flow of which is perceived by the vestibular nerve, allowing it to help the body orient itself to an upright position. The dizzy, sick feeling one gets from rolling quickly down a hill or from riding a roller coaster occurs because the rapid motion has overwhelmed the ability of the vestibular system to adjust to the pitch and yaw. The semicircular canals are adjacent to the cochlea (the head of the snail).

The key hearing element in this amazing structure is an array of about 30,000 hair cells that line the two and one-half turns of the cochlear head. Collectively, the cells, each of which has about 60 tiny hair-like projections, comprise the organ of Corti (named for the young Italian physician who first described it while working at the University of Wurzburg in 1851). The hair cells send projections to about 30,000 nerve cells that form the bundle known as the eighth cranial or auditory nerve. Vibrations arriving from the outer world press the hair cells, each of which is sensitive to a particular frequency, thus generating the electrical impulses that stream down the auditory nerve to the brain.

Physicians think of deafness broadly as arising from defects in one or more of three systems: the conductive, the sensorineural, and the central processing systems. Conductive hearing loss arises due to congenital malformations of the eardrum and middle ear, scarring from repeat infections, or any of a variety of other causes that compromise the delivery of sound waves to the cochlea. Sensorineural hearing loss is caused by defects involving the cochlea or auditory nerve. (As I discuss below, such defects are often caused by mutations in genes that are responsible for the fine structure of the cochlea.) The term "central hearing loss" encompasses all defects in the brain's ability to process signals it receives.

I have been trying to imagine deafness—profound, complete deafness. I cannot. The quieter I try to be, the louder seem the background noises (the hum of my computer, the wind at the window, the creak of the heating system). I can, I think, imagine silence. I was once at the bottom of a deep cave to which no auditory signal from the planet's surface could penetrate. The group of people I was with dispersed, shut off their lamps, and sought stillness. My eyes perceived nothing. No one made a sound, but I could still hear the rush of blood through my body. Silence was what I experienced in the fraction of a second between each beat of my heart.

Profoundly deaf people do not hear the world, but they do feel it. Their hands and feet and faces constitute a giant organ that discerns vibrations, and tiny regions in their brains, compensating for the loss of hearing, develop the ability to wring more meaning from those unending waves of pressure that strike our bodies than those of us who hear ever could. Over the years, coaches of the football team at Gallaudet, the national college for the deaf in Washington, D.C., have used drums to send signals to the quarterback. After winning a game, Gallaudet students sometimes dance to the vibrations of a cranked-up sound system. Incidentally, the Gallaudet football team gets credit for inventing the huddle, a tactic originally used to make sure players on the opposing team could not read their sign language!

Of the more than 6,000,000 Americans who have hearing loss, most are older adults who suffer mild to moderate impairment. About 10% of the total are children who were born with a serious or profound impairment. About 1 in 1,000 children is born profoundly deaf, and an equal number are born with significant hearing impairment. There are many causes of deafness and hearing impairment, including infection, trauma, side effects of certain drugs, and the conductive loss of the bones in the middle ear that is common in old age. In infants, however, deafness is mostly the result of being affected with one of an ever-growing number of genetic disorders.

Because it is relatively common (5–50 times more common than most of the genetic disorders for which we screen all infants at birth), deafness was among the first disabilities on which physicians, scientists, and social reformers focused as medicine embraced research in the 19th century. Long before European scientists rediscovered Mendel's laws of inheritance, Alexander Graham Bell studied the inheritance of deafness. In the early 1880s, he collected data on the extended families of couples in which both spouses were deaf. Bell found that in one-third of marriages in which both parties were deaf there was at least one deaf relative. In one study, he found that when a hearing individual married a deaf partner, the couple had a 20% chance of bearing a deaf child, whereas if two deaf persons married, the risk of bearing a deaf child was only 10%!

Bell knew that hearing persons who married deaf persons often had deaf relatives. From this he inferred that these hearing spouses might have a predisposition to deafness that could manifest in their children. In making this inference, Bell came within a whisker of discovering the principle of recessive inheritance. As he and his colleague, E.A. Fay, continued to study deafness, they realized that the condition was actually a large collection of different disorders. In the early 1890s, Fay concluded that if two deaf people whose deafness arose for different reasons married, the chance that they would have a deaf child was quite small. This discovery, about which he was certain by 1898, also brought him very close to grasping the concept of recessive inheritance.

Bell, whose father had spent years trying to develop a novel method of teaching communication to deaf people, began his life with the same career plan. When the deaf school he opened in Boston in 1872 failed, he turned his attention to tinkering with a device he had conceived for sending mechanized speech. In 1876, he had constructed a functional telephone. When Bell grew wealthy from his invention, he founded the Volta Bureau to conduct research on the education of deaf children and to push his view that lip reading and oral education were superior to sign language. In the last decade of the 19th century, the teaching of sign language declined drastically.

Over the last 200 years, the place of deaf persons in the hearing world has undergone profound change. The most important date in the history of efforts to help deaf persons is 1817, the year that a minister, Thomas Hopkins Gallaudet, and his French colleague, Laurent LeClerc, founded the first permanent school for the deaf in Hartford, Connecticut. Wishing to bring God's word to deaf persons, Gallaudet went to Paris, then home to the most advanced approaches to teaching the deaf. He convinced LeClerc, one of the French leaders, to immigrate to the United States and to help develop an American version of the sign language used in France. Over the next few decades, many schools like the Hartford institution were founded in the other states and territories. In 1864, President Lincoln signed a law creating the nation's first college for the deaf (then called The National College for the Deaf and Dumb) with Thomas Miner Gallaudet as its first president. In 1893, it was renamed Gallaudet University in honor of his father.

During the last third of the 19th century, deaf children (especially if they were the only affected persons in their families) were not infrequently sent to live in state-supported or privately financed institutions. Some of these schools (or asylums, as they were often called) segregated such persons; others served the deaf, the blind, those with epilepsy, and the mentally retarded. The availability of options to institutionalize deaf children suggested that society viewed them as having a very significant handicap. Several states even enacted laws (albeit never really enforced) that forbade deaf persons from marrying!

However, most deaf children were raised at home, and many (through the auspices of their older, deaf relatives) were able in time to become part of an extended deaf community. The most famous of these took root on Martha's Vineyard, off the coast of Massachusetts. In the 19th century, there were so many deaf people living on Martha's Vineyard that in some villages American Sign Language was the first language, and hearing folks had to master it if they wanted to be fully integrated into their communities. Not so many years ago, one could still find old timers who remembered summer afternoons when one could walk down a street lined with front porches filled with people drinking lemonade, and not hear a spoken word. Deaf society on Martha's Vineyard (and a few other spots in the United States) was substantially weakened when social progressives started insisting that deaf children be taken off-island to receive proper instruction in oral education.

In at least one-half of all children who are born deaf, the cause is genetic. Armed with the tools of molecular biology, scientists have identified more than 100 genes in which a mutation can be the cause of deafness. Some of these lead to syndromes in which deafness is only one of several serious problems (for example, Usher syndrome which includes blindness and Pendred syndrome which includes thyroid disorders), but in most (about two-thirds) there is no other defect. Despite the very large number of genes in which mutations can cause deafness, one in particular, called Connexin 26 (or GJB2 for gap junction protein Beta-2), accounts for more than one-half of all profound congenital deafness in many (but not all) populations, including persons of European descent.

Connexins are proteins that are a key component of gap junctions, structures that create communication channels between cells. The number 26 is scientific shorthand for the molecular weight of the protein (about 26,000 hydrogen atoms). Since it was cloned in 1992, scientists have conducted hundreds of studies on Connexin 26. In 1997, one group demonstrated that this protein was found at unusually high levels in the cochlea. This set off intensive study of its role in hereditary deafness. During the period 1997–2003, studies of deaf children throughout the world showed that in virtually every ethnic group, mutations in this gene were a major cause. Because the Connexin 26 gene is quite small (it has only two exons of which only the second acts as the template for the protein), it was quickly recognized that it might be possible to develop DNA tests to diagnose molecular deafness. It turns out that although more than 50 mutations have been found, just three—35delG in Caucasians, 167delT in Ashkenazi Jews, and 235delC in Orientals—account for most of the deafness alleles. About 1 in 30 Europeans and 1 in 25 Ashkenazi Jews are born with one of these mutations. As most forms of Connexin 26 deafness are recessive, an affected person is usually the child of two carriers who hear normally and neither of whom has a deaf relative.

Why are these connexin mutations so common in the population? Usually, as, for example, in the case of the mutation that causes sickle cell anemia that is present in about one in *ten* persons of African background, having one copy of a recessive mutation is (from evolution's perspective) beneficial. Persons with one copy of the sickle cell allele are less vulnerable to contracting malaria (the parasite lives inside red cells). So far, no one has found a plausible benefit associated with being a carrier of a Connexin 26 mutation. Walter Nance, a prominent geneticist at Virginia Commonwealth University, has come up with an intriguing alternative explanation. He points out that the populations in which Connexin 26 mutations have the highest prevalence are those in which there is a long tradition of intermarriage among deaf persons.

Over the last half century, the deaf community in the United States— the hundreds of thousands of persons who were born deaf or became deaf in early childhood and who strongly self-identify as members of a linguistically distinct minority group—have made great strides in reclaiming their right to be full participants in society. This ascendance began during World War II, when the United States army refused to induct deaf men into the military forces. Due to the labor shortage, thousands of deaf men and women were hired for jobs that they would not have been offered in peacetime. This fostered the creation of large deaf communities, for example, in Akron, Ohio, where there were many large factories devoted to the war effort. The civil rights movement of the 1960s greatly energized many young deaf persons. During the 1980s and 1990s, federal legislation ended the educational segregation of deaf children (who were among the most successful of the groups of children with learning disabilities who were mainstreamed into regular classrooms). Among the most symbolic moments in the ascendancy of deaf culture was the appointment of the first deaf president to head Gallaudet University in 1988, an event that was driven by student protest. Today, we have deaf professional athletes, a deaf Miss America (Heather Whitestone in 1994), and prominent deaf actresses such as Marlee Matlin.

When genetic chance and physical or cultural isolation create communities in which deafness is unusually common, they can be the fountainhead of new languages. Such events, rare linguistic Krakatoas, are of unsurpassed interest to those who seek to understand the nature of the universal capacity for language. In 1977, at a large school for the deaf in Nicaragua, teachers and linguists noticed that the deaf children were spontaneously developing their own sign language, replete with rules for grammar and syntax. The more recent birth of a new language among the deaf is going on right now in a culturally isolated Bedouin village in the Negev desert in Israel. The 3,500 people who live there are all the descendants of a single founder who arrived from Egypt about 200 years ago. Of their five sons, two were born deaf. Today, about 150 of the villagers are deaf. In addition to them (as was the case in Martha's Vineyard), many of their hearing relatives are learning the new language. The linguists who are studying this new sign language are especially fascinated by the speed with which rules for word order and sentence structure are emerging, and they delight in speculating how complex it may become!

Ironically, just as deaf persons have achieved a high level of integration in the United States, they are facing two technological threats that could dismember their community within a century. The first of these is cochlear implants. Developed in the early 1980s, these implantable electric devices are used to overcome problems in the middle and inner ear by directly interfacing with the auditory nerve to transmit information of sufficient sound clarity to the brain. In essence, a cochlear implant consists of a microphone, a speech processor (which arrays sound for transmission), a transmitter (which converts sound to electrical impulses), and a set of electrodes that send the impulses to the brain. Over the last 20 years, the power of cochlear implants to restore hearing has improved markedly.

Many studies have shown that if the devices are implanted early enough (preferably before age two), children who are profoundly deaf can achieve a level of perception where they function well in a hearing world. More than 70,000 persons in the world have undergone cochlear implantation, and the number of surgeries performed each year is rising rapidly.

Each time a young child receives a cochlear implant, he or she is effectively removed from membership in the Deaf (with a capital D) Community. The emigration from one community to another is less likely when the surgery is performed on adults, but it still tends to pull those with implants away from their deaf friends. One dramatic example of this is provided by Heather Whitestone, the first deaf Miss America, who a few years after she was crowned in 1994 underwent cochlear implantation and subsequently announced that she had returned to the hearing community. With the quality of the devices improving and the number of surgeries growing steadily, one can foresee a time in the near future when the vast majority of parents who give birth to deaf children will choose the operation if the device is medically appropriate. In the United States, this could sharply reduce the population of children who might otherwise have entered the Deaf Community—perhaps by more than one-half.

Not unexpectedly, there is a sharp disparity across socioeconomic groups in the frequency of cochlear implants in profoundly deaf children. In a large 1997 study, white and Asian children in the United States were more than three times more likely to receive implants than were Hispanic children. Black children with comparable impairment were ten times less likely to obtain the devices. Children who received cochlear implants were far more likely to live in communities with high median incomes. This finding is just one of hundreds one could report about differential access to health services being determined by economic status. It also suggests that in coming decades the Deaf Community could have a proportionately higher membership of persons of color.

A new emphasis on prompt diagnosis of deafness in infants through newborn screening, combined with the growing ability to provide a precise molecular diagnosis for the majority of hereditary deafness, poses a second threat to the future of the Deaf Community. Unlike the situation a decade ago, today most children with severely impaired hearing are identified in the first weeks of life. In the vast majority of cases, their normally hearing parents want to know the recurrence risk for the next pregnancy.

Since most congenital hearing loss is due to recessive disorders (creating a one-in-four risk in each pregnancy), most at-risk couples have had one or more hearing children before giving birth to their deaf child. Many of those couples simply choose not to have more children. Some (in part because of the growing efficacy of cochlear implants) may choose to continue childbearing without making use of genetic testing. For those who want more children, but do not want to bear a deaf child, there are additional options, including adoption, preimplantation genetic diagnosis (genetic testing on embryos created by in vitro fertilization permits couples to implant only those who will not be born deaf), and (rarely) prenatal diagnosis and selective abortion. Although there is little evidence to document the trend, it may be that the number of children born with congenital deafness in the United States is already declining. That fact, combined with the soaring use of cochlear implants, assures that over the next few decades many fewer children in the United States will grow up with sign language as their major form of communication.

The vital center of the Deaf Community is composed of deaf men and women in their 20s and 30s who use sign language, abjure cochlear implants, and would be happy to have or, in some cases, would overtly prefer to become, parents of deaf children. For example, at a Deaf Nation conference in England in 1987, of 87 delegates who were asked, 14 answered a questionnaire saying that they would be interested in prenatal diagnosis. Of those, 4 said they would like to use it to have a deaf child. Just as hearing parents could use genetic tests to avoid the conception or birth of deaf children, deaf couples could use such tests to ensure the birth of a child who is deaf. In 2002, a deaf couple in Australia used preimplantation diagnosis to avoid having a deaf child. Regulatory authorities in the state of Victoria have stated that they would not permit the same technology to be used to select for a deaf baby. The few studies that have been done indicate that only a small percentage of deaf couples would terminate a pregnancy because the fetus would be born as a hearing child. But those few that would do so challenge the traditional use of prenatal diagnosis, which is to give couples the option to avoid the birth of a child with a serious disease or disability. Of course, they also challenge the definition of disability.

For four decades, the vast majority of human geneticists and obstetricians have opposed assisting couples in screening and terminating pregnancies on the basis of sex, arguing that gender is not an abnormality. Few (if any) would assist deaf couples in the abortion of hearing children. Is it ethically permissible for deaf couples to knowingly avoid the birth of hearing children and knowingly seek the birth of deaf children? For some people, the answer to this question is determined by the technology that is used. Many people would be considerably less uncomfortable with the use of in vitro fertilization, preimplantation genetic testing, and transfer of embryos destined for deafness than with prenatal diagnosis and abortion of hearing fetuses. If a woman has the moral right to decide whether or not to continue any pregnancy for any reason up until a certain point in her pregnancy (essentially, the definition of her current legal right), does it not follow that she has a moral right to decide whether or not to become pregnant with any embryo for any reason? Should she be free to terminate a pregnancy for any reason she wants?

In seeking to bear deaf children, young adults in the Deaf Community have challenged mainstream conceptions of normalcy. Imagine yourself born deaf, attending a deaf preschool program where you learned sign language, schooled for twelve years in classes filled only with other deaf children, attending a college for the deaf, and married to a deaf spouse. Is it really so hard to imagine wanting to have a child who will live in the same world as you do? Is it not easy to understand how young deaf adults can feel sharply at odds with the hearing parents of young deaf children over what world would be better for those children?

From time to time, technological advances in medicine pose immense threats to certain groups. In the United Kingdom, prenatal screening and selective abortion have sharply reduced the live births of children with spina bifida. There is a similar trend in regard to fetuses with Down syndrome. In the United States, the decline in live births is less than in the United Kingdom, but it is still notable. Similarly, during the next several decades there will be steadily fewer children born with profound deafness. Most deaf newborns will receive cochlear implants by the age of 2.

Walter Nance, the aforementioned expert in the genetics of deafness, has worried that we are unintentionally using new medical tools to foster a cultural genocide (ethnocide) in which a community of able, but different, individuals disappears from the societal fabric. Is that the right view? Or, in contrast, should we regard the rapid growth of cochlear implant surgery as an effort to combat a serious childhood disorder, much as we struggle to treat juvenile diabetes? Of course, there is value in both perspectives. It seems to me that the only choice is to pursue a dual policy. We must continue our efforts to maximize the opportunities for those born with serious hearing impairments, even as we provide couples the option to avoid becoming parents of children with such disorders.

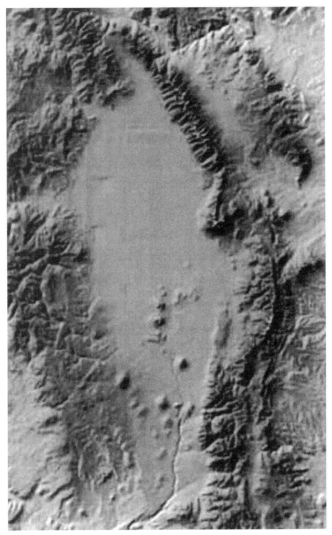

A topographic map of the San Luis valley in southern Colorado.

9

San Luis Valley Syndrome

The San Luis valley in southern Colorado is one of the largest high desert valleys on earth. About the size of Massachusetts, this exquisitely beautiful alpine land is bounded by two great mountain ranges, the Sangre de Christo to the east and the San Juan to the west. To the east, at over 14,000 feet, Mount Blanca dominates the terrain. The mountain ranges generate local weather patterns that include spectacular cloud formations and the most sharply defined triple rainbows I have ever seen. Great Sand Dunes National Park, a spectacular pile of sand laid down by the erosive effects of thousands of years of wind roaring through a narrow channel of mountains, lies in the northwest corner of the valley. The Rio Grande, which has played a crucial role in the valley's history, begins as a tiny stream in the San Juan range from which it runs several hundred miles south through New Mexico.

The San Luis valley is part of a vast tract of land that the United States acquired from Mexico by the Treaty of Guadalupe Hidalgo, which ended the Mexican-American War in 1848. In exchange for $15,000,000 and the assumption by the United States of all liabilities arising from the claims of U.S. citizens against it, Mexico gave up its claim on lands north of the Rio Grande, some 500,000 square miles, including all of California (where gold was discovered the following year, propelling it to statehood in 1851). Initially, the San Luis valley was part of the New Mexico Territory that was created in 1850, but in 1861, the year Congress established the Territory of Colorado, ambitious settlers, using what some people in New Mexico still call questionable legal maneuvers, managed to fold most of the valley into the newer territory.

For several millennia before the arrival of the Spanish conquistadores, various Native American peoples periodically occupied the valley, people who left so sparse an archaeological record that we can do little more than speculate about their lives. For about three centuries (1600–1900), the valley was often home to the Ute Indians, a nomadic people who had much interaction with New Spain (the forerunner of Mexico), which laid claim to the territory in the early 1600s. The first permanent (nonnative) settlers in the San Luis valley were five mestizo (individuals of mixed European and Indian heritage) families who migrated about 100 miles north from the Taos valley in what is now New Mexico. They arrived in 1851.

Despite its beauty, Costilla County, which encompasses much of the valley, is today sparsely populated. There was a big growth spurt in the 1870s when gold and silver were discovered in the San Juan mountains, growth that was reinforced when the railroad reached the central town of Alamosa in 1878. But as mining prospects faded, people moved on. The county population reached a high of about 7,500 in 1940; in 1999, it was 3,600. Many of its citizens can trace their history to one or more of the original families, a key element of this genetic tale.

Sometime in the 17th or 18th century, an ancestor of one of the first families to settle in the San Luis valley was born with a rearrangement of the DNA in one of his or her chromosomes. During the formation of the egg or sperm that united to create that ancestor, the usually flawless enzymatic machinery that copies DNA erred. The copying error caused a large chunk of the genetic material on chromosome 8 to be cut in two places and then inverted (flipped 180 degrees) before it was chemically rejoined to the rest of the chromosome. The mistake did not harm the germ cell. All the genes were still present, albeit in a new alignment, and the child who was born with a rearranged chromosome 8 was healthy.

That child grew up, married, and had a family. Given the era, the Hispanic Catholic culture, and the importance of children to an agricultural economy, it was quite likely a large family. Two or three centuries ago, it would not have been uncommon for a woman in that culture to bear ten or more children. Nor would it have been uncommon for several of them to die in infancy, usually of infections. In this family, however, a family in which one parent carried the inverted 8 chromosome, if one or more of the children died in early childhood, it might well have been for a unique reason. They may have died of a disorder that had never before struck down a human being. The parents could not know of their risk, but each time that the woman conceived, there was about a 6% chance that the infant would be born with severe abnormalities due to having inherited an abnormal amount of genetic material found on chromosome 8, a condition that is today called Recombinant 8 or San Luis valley syndrome. Important to this tale is that each of the children in this family was born with nearly a 50% chance of having an exact copy of the inverted 8 chromosome that was in the germ cells of one of the parents.

As is often the case with severe birth defects caused by abnormal chromosomes, the discovery of recombinant 8 (rec8) syndrome arose out of the study of an infant who was born with an unusual constellation of physical abnormalities. The initial case report (as they are called) was made in 1975 of a little girl who died of heart failure and multiple other problems at the age of six weeks. Careful study of the chromosomes of the deceased infant found that she had two marked abnormalities of one of her number 8 chromosomes. The family lived in the San Luis valley.

All humans are born with 23 pairs of chromosomes (one set from each parent): 22 pairs of autosomes and one pair of sex chromosomes (women are 46, XX, and men are 46, XY). The chromosomes are numbered in descending order of their size as seen under the light microscope, with number 1 being the largest. The X and the much smaller Y are not part of the numbering scheme. This identification system was developed in the 1960s. The scientists working in this new field agreed that the study of human chromosomes, each of which has a fairly distinctive shape due to the presence of long and short arms attached to a centromere, would be enhanced if the letter p was universally used to denote the short arm and the letter q to denote the long arm. Thus was born a scientific shorthand; when human geneticists read that a child has 4p– syndrome, they know that he or she is missing some of the genes on the short arm of chromosome number 4. An agreement to name the human chromosomes according to specific rules was the beginning of a mapping effort that culminated 40 years later in the completion of the human genome project.

In the early 1970s, several different teams of scientists discovered that individual chromosomes had distinctive banding patterns when they were treated with certain dyes. This permitted scientists to be more precise in their analysis of chromosomal aberrations. By agreeing on a common system of numbers to designate the bands that were consistently produced by treatment with the dyes, cytogeneticists (scientists who specialize in chromosomes) could overlay several hundred place names on the human karyotype.

The little girl who died of rec8 syndrome in 1975 had both a duplication of a significant chunk of the long (q) arm of chromosome 8 and a deletion of part of the short (p) arm. In cytogenetic parlance, her condition is called: rec(8)dup(8q)inv(8)(p23.1q22.1). The term, p23.1q22.1, denotes the material in the segment of the chromosome that was inverted (the two points at which breaks occurred to permit the inversion). A human geneticist reading this paragraph immediately recognizes that the chromosomal material on the short arm distal to band number p23.1 is missing and that the material distal to 8q22 is duplicated. One of that girl's parents had an inverted chromosome 8 (a pericentric inversion, meaning that the flipped material included the centromere).

Although people born with the inversion are themselves normal, the oddly shaped chromosome is unstable. During the formation of germ cells (meiosis), when the DNA undergoes a complicated process of doubling and then reduction to the haploid (one copy rather than two of each chromosome) state, the inverted chromosome 8 is at much higher than normal risk to undergo a serious copying error. Persons who carry an inverted 8 have about a 6% chance that a sperm or egg formed by their germ cells will include a chromosome 8 that has been duplicated in part and deleted in part. This statistical risk was determined by studying the reproductive history of families in which one parent is a carrier of an inverted but otherwise normal 8. No one knows exactly why the chance that a person carrying an inverted chromosome 8 will become the parent of a child with rec8 syndrome is 6% (as opposed to, say, 30%).

It is intuitively obvious that persons whose cells are missing big chunks of DNA (and therefore, genes) will have significant abnormalities. The same, however, is also true for people born with extra chunks of DNA. A child born with rec8 syndrome has three copies (instead of two) of scores of genes on chromosome 8. This "dosage effect" can be very harmful. A better-known example of this is manifest in people who have Down syndrome. That condition, one of the most common of all chromosomal abnormalities, is caused by the presence of an extra chromosome 21. Number 21 is quite tiny compared to 8. The person with Down syndrome has three copies of all the genes on chromosome 21, but because it is so small, the dosage effect is somewhat less severe. To give some perspective on this, no fetus conceived with a full extra copy of the large chromosomes (conditions called trisomy 1, 2, or 3) has ever been born alive. These trisomies are only found in the tissue of spontaneously aborted fetuses.

As is always the case with new discoveries in medicine, the first case report of a child to die with rec8 syndrome stimulated physician-scientists to look for other children with similar physical problems who might have the same underlying cause. Since the little girl who was first described with rec8 syndrome had died because of severe developmental abnormalities of the heart, researchers focused their search for other cases on infants with a similar problem. New cases were found quickly. In 1981, a team led by Arthur Robinson, a prominent pediatric geneticist at the University of Colorado, reported that he and his colleagues had diagnosed the rec8 syndrome in eight children from seven families. All the children had similar problems: an unusual facial appearance, severe developmental delay, seizures, and congenital heart disease. Given the magnitude of the chromosomal duplication, this constellation of findings was not surprising. However, one aspect of the report was. All the children were of Hispanic origin and all had been born to families in the San Luis valley!

By 1987, the group led by Robinson had diagnosed rec8 syndrome in 26 children and collected case reports on 5 others. All 31 children were of Hispanic background, and all the families could be linked historically to just three kindreds, the ancestors of whom had lived in the San Luis valley in the 19th century! The evidence of a founder effect—that all the children with rec8 syndrome were descendants of a single individual who was born several centuries ago with an inversion of chromosome 8—was irrefutable!

By 1990, just 15 years after the first case of rec8 syndrome was reported, pathologists, pediatricians, and human geneticists were calling the new condition San Luis valley syndrome. Every child who had ever been diagnosed had roots in the valley. Importantly, the scientists and clinicians now grasped that the condition was not as uncommon as they at first thought. They also realized that close study of the abnormalities combined with molecular analysis could yield important new knowledge about how the human heart develops and which genes play a major role in forming its structures.

This has proved to be true. As the number of cases that has been studied has grown, it has become apparent that nearly 95% of all children born with rec8 syndrome have heart defects. About 55% have conotruncal defects (those involving the formation of the channels that permit blood to flow out of the heart) and about 40% have a condition called tetralogy of Fallot (named for Etienne-Louis Arthur Fallot, the Marseille physician who first described it). As its name indicates, tetralogy of Fallot involves four separate abnormalities: (1) obstruction of blood flow from the right side of the heart to the lungs, (2) a hole in the wall between the right and left sides of the heart, (3) a faulty positioning of the aorta, and (4) an enlarged right heart chamber.

In the United States, about 1 in 200 children is born with a heart defect, but no other problems. In this group, conotruncal defects and the tetralogy of Fallot are much less common than among kids with rec8 syndrome. Because children with other kinds of chromosomal 8 abnormalities are not nearly so likely to have these particular heart defects as are those with rec8 syndrome, the scientists concluded that certain genes on chromosome 8 play a key role in ensuring the normal development of the heart. However, at that time they were unable to attribute the cause to duplicated genes, missing genes, or genes that were interrupted by the recombination event. A steadily growing number of case reports concerning children who are born missing a small part of the short arm of chromosome 8 and with heart defects does make it quite likely that it is the loss of genes in that region that causes the heart defects in San Luis valley syndrome.

It is often the case that the earliest clinical descriptions of a genetic disorder derive from patients who constitute the most extreme severity of expression. Today, three decades after the first description of a child with rec8 syndrome, we know that there is some variation to the natural history of the disorder. Although virtually all the children have heart defects, and all have mental retardation and mild abnormalities of the face, the severity of each case varies. In 1993, the team at the University of Colorado School of Medicine described 42 patients. Although one child had died at age 5 days, another was alive at 23 years of age. The age of death has been mainly determined by the severity of the congenital heart defect, but the socioeconomic status of the family is also a factor. All children with rec8 syndrome require massive amounts of care. Elements as disparate as the geographic location, the skill of the primary care team, the educational level and the wealth or poverty of the parent(s), and the existence of other family stresses can all affect the life expectancy and quality of life of children with severe birth defects. The words of a woman (I will call her Maria) who is the mother of a teenage boy with rec8 syndrome paint a vivid picture.

Maria remembers that her son (whom I will call Pedro) underwent heart surgery before he was three months old and that he was in the hospital for three months. When he came home he needed to be on supplemental oxygen and had weekly visits to the doctors for more than a year. After Pedro was born, the parents had to reorganize their lives. In the words of Pedro's older sister, "Everyday requires constant care, preparing special meals, administering medications, clothing, changing, bathing, and finding the little things in life that keep the child content."

In families caring for children with San Luis valley syndrome, one parent is always at home. The families almost never go on vacations because they feel the need to be near their doctors. Among so many other problems, affected children have weak immune systems, and a common cold can easily lead to a lengthy hospitalization. By the time he was two, it was clear that Pedro had significant developmental delay. He required a complex, time-consuming early intervention program that promised only very modest gains. Now in his late teens, Pedro cannot read or speak. He goes to school but spends most of his day in gym and art classes. He has learned to communicate some of his needs and interests by pointing, but only caretakers who know him well can decode his communication. As Maria puts it, "He has trained us well."

Because the syndrome occurs only in the children of persons of Hispanic descent who trace their ancestry to the San Luis valley and who carry an inverted 8 chromosome, one would think it a straightforward task to identify persons at risk for bearing affected children to warn them before they do so. Because the disorder is now well known among residents of the valley, one might guess that a family history of a child who died young or who had to undergo surgery for heart disease would trigger immediate concern among relatives who are thinking about having a family. However, as is often the case with familial disorders, especially when the individual risk is relatively low, a mixture of misunderstanding, denial, guilt, and acceptance complicates efforts to warn.

If the risk to each pregnancy in which one of the parents is a carrier of the "balanced" form of the unusual chromosome is only 6%, then many families who are at risk will, because they have not been exposed to an affected relative, have no basis from which they can infer their risk of having an affected child. It would be possible to conduct an extensive survey within the valley and then perform chromosome studies on every person who reports having a relative who died in infancy, had heart surgery in childhood, or has mental retardation. This would be quite expensive. In addition, such a public health approach to this disorder would raise a troubling set of ethical questions. Perhaps such risk screening would be viewed as a modern form of eugenics.

Most important is to ask about the consequences for the individual of learning that he or she carries the inverted chromosome 8. Would such a discovery frighten and stigmatize teenagers? Would their peers regard them as poor prospects as future husbands and wives? Could the diagnosis of being a carrier lead to social isolation and/or perhaps clinical depression? Among adult couples who have not had an affected child, but who are discovered to be at risk, what options can be offered? Most of the Hispanic residents of the San Luis valley are devout Catholics, and many would probably have great difficulty in using amniocentesis to test the fetus and even greater difficulty in opting for selective abortion. The people probably have almost no familiarity with options such as sperm or egg donation. Preimplantation chromosome diagnosis might be a more welcome ethical choice than amniocentesis and selective abortion, but it is very expensive, and many people would still abjure it as contrary to the teachings of their church.

Gene and chromosomal variants flow across the planet, reflecting the movement of individuals and peoples. Over the last decade, our ability to trace gene variation has revolutionized our understanding of the history of human migration patterns. The San Luis valley is not the geographically isolated region it was a century ago. Each decade, more young people choose to seek their fortunes elsewhere. Doubtless, there are hundreds of persons alive today who are unaware that they carry the recombinant 8 chromosome which puts them at risk for having a child with severe birth defects. As they leave the valley for college or military service or adventure, they carry that chromosome with them. Already, affected children have been born in Los Angeles and New Orleans.

Today, the recombinant 8 chromosome may have a competitor for the eponym, San Luis valley syndrome. In 2001, a team of clinical researchers at the University of New Mexico Health Science Center published the results of their extensive study of a disease called oculopharyngeal muscular dystrophy (OPMD). Affected individuals develop both eye muscle and throat muscle weakness early in life and weakness of the extremities later in life. Although they live a normal life span, OPMD patients have many problems related to difficulties in swallowing and general muscular weakness. This dominantly inherited disease is caused by an expansion of a trinucleotide (three DNA bases, GCG) repeat within a gene called PABP2 which disrupts its protein. This is the same sort of genetic error that causes Huntington's disease. The 216 patients that the research team identified belong to 39 extended families. All are Hispanic and all trace their family history to northern New Mexico! In Hispanics in the United States, OPMD almost certainly arose from a founder. However, it is possible that the founder will be ultimately traced deeper in time to Spain before colonialism. A large number of patients who apparently have the same disease have been diagnosed in Uruguay, all of whom are descendants of families who came from the Canary Islands, an area colonized by Spain in the 16th century.

OTHER FOUNDER EFFECTS

Not all chromosomal oddities that arise in particular ethnic groups cause disability or disease. However, it is almost always the case that the variant can be used to help reconstruct or confirm the history of the migrations of the group. During the 1980s a chromosomal study of the Khanty, a people who live along the lower Ob River in West Siberia, found that more than 20% of the men carried a large deletion of genetic material on their Y chromosomes. Detailed anthropological and genetic study of the group led scientists to conclude that the variant, which appears to be benign, is the result of a process called *genetic drift*. This is essentially a random process whereby a neutral variant (not buffeted by the forces of natural selection) that arose in a single person in the distant past achieves a high frequency simply by chance. One might liken it to a particularly good run of luck with coin tosses over time.

A similar study of the frequency of a variant of the Y chromosome (an inversion) in a Muslim Indian community living in the Transvaal Province of South Africa yielded surprising results about its migration history. A study of 72 normal adult men in the community found that 22 (30%) had the inverted Y. All the men with the inverted Y belonged to families that could trace their ancestry to a few villages in the city of Surat in Gujerat Province in India. In this community, random genetic drift was probably reinforced by strict endogamy (marrying within a specific group). It is still socially unacceptable to marry outside one's religious or linguistic group.

As we improve our tools for searching for really small chromosomal inversions, translocations, and deletions, we will certainly find more. As with the Gujerati, some of these will merely be markers of ethnicity. For example, in the early 1990s, an inversion of part of chromosome 11 was found among a branch of the endogamous (intermarrying) Hutterite people in Canada. In other cases, variant findings will explain disease that correlates strongly with ethnicity. In 2003, researchers in Arizona showed that the comparatively high frequency of oculocutaneous albinism (affected persons lack skin pigment and have severe visual problems) among the Navajo was due to a 122,000-base pair deletion in the "P" gene. About 1 in 22 Navajo carries this deletion, but it has not been found in other Native American groups. This disease (called OCA2 for short) is recessive; only about 1 in 1,500 Navajo is born with it.

One of my favorite stories about using genetic variation to reconstruct population history involves the European Roma/Gypsies. Believed to have arisen originally as a single small founder population, the Roma probably have split apart several times during the last 1,000 years. Given that the Roma are a wandering people who have not kept written records, their origin is shrouded in mystery. In 2004, scientists in Australia capitalized on the existence of disease-causing mutations to establish that the Roma almost certainly derive from the region that today is Pakistan. By studying the distribution of several small variations in the genome, the researchers were able to estimate that the Gypsy diaspora began 32–40 generations ago, and that two further divisions occurred about 16–25 generations ago.

The discovery of chromosomal variants can also provide scientists with hints about the location of disease-causing genes. In recent years, the long search for the causes of male infertility has led scientists to look for chromosomal variants. The largest study of the relationship between chromosomal variation and infertility was conducted in Japan in 2000. In analyzing the chromosomes of 1,790 infertile men, the scientists found abnormalities in 225. The most frequent diagnosis was Klinefelter syndrome (affected men have an extra X chromosome) with 64 cases, but there were also 30 cases of variants involving chromosome 1. In 2004, scientists in Denmark studied the chromosomes of 464 men with unexplained infertility and compared the findings to a similar study in a control group. They found that infertile men were twice as likely as fertile men to have a variant on chromosome 1. The largest number of variants occurred in the region called 1q21, leading the scientists to speculate that it is the location of a gene which, if mutated, causes reduced fertility.

One of the hardest clinical problems in studying newly discovered chromosomal variants is whether or not to designate them as benign or dangerous. In 1998, a team at a UK hospital reported on seven separate families in which they had found some members to have a duplication of some of the material on the short arm of chromosome 8, but all of whom appeared normal. This is most probably a benign variant that has spread by genetic drift in the Wessex region of England. Unfortunately, it is very difficult to be able to state unequivocally that the duplication causes no health risks. During the early 1990s, one research group described a variant of the tiny end of chromosome 21 that it thought was associated with mental retardation. Later reports disputed that, asserting that this so-called "Christchurch" variant was harmless.

In the not too distant future, the study of cytogenetics will belong to the historians of biology and medicine. Scientists will use the unique signatures in DNA itself to trace the origin of diseases that arose by founder effect or to trace the movements of small, highly defined human groups. There will be more stories like San Luis valley syndrome. The intractable question of how best to use this knowledge to find and warn carriers will be with us for decades.

Alain Fischer and Marina Cavazzano-Calvo announcing success in their gene therapy research with SCID. (Reprinted, with permission of Macmillan Publishers Ltd., from Check 2002.)

10

Severe Combined Immune Deficiency

For more than a billion years, cellular life, buffeted by countless mutations and shaped by the myriad forces of natural selection, has drifted purposelessly along the currents of time, yet achieving ever greater complexity. The odds that any particular species among the millions that inhabit the earth should be alive today are so astronomical that they offer no help in understanding its presence. We, and those who inhabit the planet with us, are here for only one of two reasons: because a Supreme Deity wanted it that way or due to chemical capriciousness that may be beyond the neurological limits of our comprehension.

There is, however, one fact about human evolutionary history of which we can be certain. We would not be here today, tending our children, writing books, mowing lawns, building hospitals, fighting wars, sending spacecraft to Saturn, and doing the millions of other activities that are within humanity's repertoire, if we had not developed a highly effective system for defending ourselves against the countless, invisible, ubiquitous hordes of small invaders (viruses, bacteria, parasites, and the like) that have evolved with a genetic program to seek entry into our bodies, dwell within us, and live well by using or consuming bits of us.

Few systems work as flawlessly or as unobtrusively as does the human immune system. During every moment of human life, the T cells (which are dedicated to defeating invaders that get inside of cells) and B cells (which create and deploy a sort of Star Wars shield of antibodies programmed to kill viruses and bacteria) are alertly standing guard, monitoring for foreign invaders, totally committed to sacrifice their existence in repelling them, and ready to mobilize legions of their fellow cells to defend the homeland. These defensive armies are the product of an intricate developmental pattern that originates in the yolk sac, begins to differentiate in the fetal liver, matures further in the thymus and bone marrow, and reaches full fighting strength after passing through the lymphoid tissue scattered strategically throughout the body. With every breath we take, untold numbers of T cells and B cells silently and efficiently eliminate viruses and bacteria that have breached the outer ramparts of the body's fortress. Until very recently—effective antibiotics first appeared less than a century ago—our immune system was all we had to protect us against the world of microbes.

Not surprisingly, a critical failure in one of the genes that code for the proteins necessary to have a well-organized army of T cells and navy of B cells poses a supreme threat to survival. A failed immune system is of as little defensive value to us as were the undisciplined mercenaries in the depleted legions of 5th-century Rome before the advancing cavalry from the steppes of Asia. Death comes quickly if your immune system is badly flawed. It comes more slowly if it is only weakened. Indeed, one of the theories offered to explain the natural limits on life span is the declining strength of our immune defenses after six or seven decades. There is a good reason why physicians used to call pneumonia "the old man's friend." Infections that we usually can fight off in midlife often take elderly people despite antibiotic therapy.

There are several dozen inherited immunodeficiency disorders, each caused by the failure of one of many different genes, most of which code for proteins involved in making T cells. All are rare—so rare that, once her training at a big medical center is complete, the average pediatrician never again sees a single patient with one of them. Nevertheless, these diseases are exceedingly important, because as we try help the children afflicted with them, we learn a great deal about the nature of the immune system. They are also extremely interesting for another reason: Genetically determined immune disorders are among the leading candidates for effective use of gene therapy. In particular, prospects for the future of gene therapy will be significantly influenced by the struggle to cure a disorder called X-linked severe combined immune deficiency (SCID).

X-linked SCID is the most common of the uniformly rare genetic immune disorders, affecting about 1 in every 25,000 boys. It affects only boys because the underlying defect is a mutation in a gene on the long arm of the X chromosome, and the Y chromosome does not carry a normal version of the gene to compensate. Fortunately, almost all girls with a mutation on one X chromosome have a normal counterpart, which is enough to provide them with a healthy defense system. However, in each of her pregnancies, a woman who carries a copy of the mutated gene on one of her Xs has a one-in-four risk of having an affected son (a one-in-two chance of having a son multiplied by the one-in-two chance that he inherited the X with the mutation).

When a family history for the disorder is known, it is possible to offer a genetic test to a woman who is at risk to see if she inherited the SCID mutation. Women who learn they are carriers and who wish to avoid the birth of an affected son can choose the option of prenatal diagnosis and selective abortion of affected fetuses. It is also possible to offer a technique called preimplantation genetic testing that would permit them to bear only girls, thus avoiding the tragic choice of aborting a wanted pregnancy. Given the small size of families during the last 50 years, the rarity of the disorder, and the fact that, until recently, affected babies often died quickly and without a correct diagnosis, many women who are at risk for bearing affected sons may have no hint of that fact. In addition, an unknown number of cases are caused by new mutations.

The causative mutations are found in a gene called the Interleukin-2 receptor gamma chain gene (IL2RGC) that codes for a protein which is part of a molecule that is absolutely necessary for the maturation of active T and B cells. The molecule is a receptor that sits on the surface of the developing immune cells and permits them to take up other chemicals which act as growth factors. Infants born with this X-linked SCID have few T cells and produce no antibodies. Although experts have shown that, depending on the particular mutation, some cases of SCID are more severe than others, almost all affected boys are ill nearly from birth with recurring infections.

These are not merely the normal viral illnesses of childhood. Children with SCID are at high risk for bacterial pneumonia, meningitis, and sepsis. They regularly become infected with organisms that fail to disturb children with normal immunity. Among the most severe is pneumonia caused by an organism called *Pneumocystis carinii*, the same bug that has killed so many patients with AIDS. Infection with chickenpox is potentially fatal. Because the children lack the defenses to contain the varicella virus, it can spread throughout their lungs and even the brain. Risk from other common viruses such as those that cause the flu, cold sores, and diarrhea is also very high. In fact, many children with SCID suffer from chronic diarrhea because so many foreign organisms have taken up residence in their intestines. The children should not be given live vaccines such as for chickenpox; in them such vaccines can cause the very diseases they are designed to prevent.

Untreated, most children with X-linked SCID would not survive beyond age two. Fortunately, SCID is a disease involving cells that arise in the bone marrow. Thus, when a perfectly matched donor is available, a bone marrow transplant cures the affected child in about 80–90% of cases. Unfortunately, because of the complex nature of our immune defense system, only about 20% of patients have relatives so closely matched that a bone marrow transplant is feasible. Although the National Marrow Donor Program has more than 4.5 million names of potential donors, the odds of finding a match among strangers through its registry are relatively low. The immune system is so genetically complex that it is rare to find matches among non-relatives. Thus, for the 80% of patients without a donor, the best hope for cure is gene therapy—the molecular transplantation of a healthy version of the SCID gene.

Despite its rarity, X-linked SCID is relatively familiar to the public, largely because of a dramatic attempt more than 20 years ago to raise an affected child in a sterile environment for as long as possible. The famous "Bubble Boy" who lived near Houston (and about whom a television movie was later made) spent a decade in a clear plastic sterile chamber, while his parents waited and prayed for a cure. His life ended when he decided he wanted to take the chance of cure by accepting a bone marrow transplant from his sister. The heroic effort failed; he died from a massive viral infection that probably traveled with the donated bone marrow into his body. The overall success rate with such operations was considerably less back then than it is today.

Although every nucleated cell contains the IL2RGC gene, X-linked SCID can be thought of as a disease of the bone marrow, the tissue which contains the stem cells that produce the billions of new recruits that the armed forces of immunity unendingly demand. This is why the disease has long been of interest to gene therapists. It is relatively easy to remove bone marrow cells from patients and to isolate the stem cells. In addition, it has become possible to isolate and transfer normal IL2RGC DNA into those cells and then infuse them (by a simple intravenous line) into the patient. Miraculously, some of the cells that return by IV find their way back home to the bone marrow. If enough genetically engineered stem cells resume their duties, as they repeatedly divide (which is exactly what stem cells are programmed to do), they create a new army of healthy immune cells in which there is a normal copy of the key gene.

But how can genes, tiny bits of DNA, be taken up by bone marrow cells in a plastic flask and incorporated into a functional spot in their nuclei? The answer lies in genetic engineering. For somatic cell gene therapy to work, genetic engineers must develop a method to attach a normal copy of the gene of interest to a virus (called a vector) that has the capacity to penetrate human cells, enter their nuclei, and insert into their DNA. Scientists have been trying to develop highly effective viral vectors for years, with mixed results. Ironically, in attempting to cure SCID, they used a virus to invade the same cells that they wished to render competent so that the patient would make T and B cells to prevent such viral invasions!

The first impressive success in human gene therapy in SCID occurred at the Necker Hospital for Sick Children in Paris during 1999–2000, a time when the field was at its nadir. Less than a year earlier, a young man named Jesse Gelsinger, who was burdened with a relatively mild form of a different genetic disorder, had died of liver failure after agreeing to undergo gene therapy as part of a research project at the University of Pennsylvania. Subsequent investigation of the case led to censure of the head researcher on several grounds, not the least of which was that he had an ownership interest in a gene therapy company. This tragic occurrence reverberated throughout the research community; for a short time, the National Institutes of Health suspended all human gene therapy research conducted with federal funds in the United States.

The team at the Necker Hospital is led by Alain Fischer, who began his research on gene therapy for X-linked SCID in 1993, just after the culprit gene was identified. He spent six years studying the therapy in mice with a similar form of the disease before trying to treat children. In April of 2000, Dr. Fischer reported that his team had performed gene therapy on two infant boys. Three months later, the little patients were able to leave their protected hospital environments for home. Seven months after that, blood tests showed that these boys had adequate numbers of normally functioning T and B cells. During 2000, Fischer's team treated eight more young boys with the disorder. After just a few months, all but one had made sufficient progress to enable them to leave the hospital's protective cocoon. Similar therapeutic research efforts were quickly initiated in the United States and several other countries. The impact on the understandably desperate patient community was profound. Parents of affected children who lacked matched donors clamored for gene therapy. Gene therapy researchers were euphoric; the triumph in France was a new beginning. But the euphoria was short-lived.

In the autumn of 2002, the team at Necker Hospital discovered that one of the nine children who had apparently been cured of X-linked SCID with gene therapy had developed a form of T-cell leukemia called T-cell acute lymphoblastic leukemia (ALL) about 30 months after treatment. In essence, the T cells in this child underwent a population explosion. Fortunately, the little boy responded to standard chemotherapy for T-cell leukemia, but the event cast a pall over both the research and the patient communities. The same question now haunted parents and physicians alike. Would more of the treated children develop T-cell leukemia? Because two relatives of the child with leukemia had also developed cancer at an unusually early age, many held out hope that this was an isolated incident, a fluke. But T-cell leukemia is sufficiently rare that a second case would unquestionably point to the gene therapy as the cause.

The second case of T-cell ALL was diagnosed four months after the first. In the United States, the FDA immediately stopped 27 gene therapy trials involving several hundred patients, all of whom were to be treated with retrovirus vectors similar to the one used at Necker Hospital to deliver the genes to the bone marrow cells. Similar actions were taken throughout Europe. Other gene therapy trials that did not involve manipulation of stem cells were allowed to continue because they were viewed as less risky.

What caused leukemia in these children? When researchers use viruses to carry new genes into bone marrow cells, they cannot dictate where in the cell's genome the virus will insert itself. For reasons that are not fully understood, some stretches of DNA are more receptive to insertion than are others. Careful study of the DNA in the leukemia cells of the two boys revealed that in both cases the viral vector integrated close to a gene called LMO2, a known oncogene (one that influences cell growth and division and which helps prevent cancer). LMO2, which is located on the short arm of chromosome 11, codes for a protein that plays its most important role early in the development of the cells that eventually become immune cells, red blood cells, cells that line the vessels, and even nerve cells. Studies in mice have shown that if LMO2 does not shut down (stop working) when it is supposed to, the mice develop T-cell leukemia.

For statistical reasons, there are grounds to suspect that in a few of the millions of cells used in the gene therapy for X-linked SCID, the virus inserted in or near the promoter region of LMO2 in a way that reactivated it. Each child received about 90 million cells that had been infected with the virus carrying the normal IL2RG gene. Assuming that there are about 30,000 genes in the human genome and that the virus inserts randomly, it is certainly plausible that among the millions of insertion events, at least 100 resulted in cells in which LMO2 was reactivated. Each of those cells can be thought of as posing a risk for T-cell leukemia. Such cells constitute a double whammy to the child's bone marrow. The successful placement of IL2RG caused T-cell proliferation; the inadvertent reactivation of LMO2 blocked their development. This left a growing reservoir of immature T cells. Such cells are vulnerable to other mutations that, as they accumulate, can result in their rapid expansion in number. This is the essence of leukemia—a stampede of white cells.

During 2002–2003, Dr. Fischer's team at Necker worked intensively to modify the gene therapy protocols to reduce the risk. There was good news early in 2004 when studies of leukemias induced in mice by gene therapy with retroviruses like the one used in humans suggested that the problems in SCID might be unique to that disease. This meant that other types of gene therapy might not carry the risk. In May, the French health agency charged with oversight of gene therapy authorized the resumption of gene therapy for X-linked SCID. The new work was to be conducted with a revised protocol which reduces the number of genetically engineered cells that could be transferred into the patients and forbids therapy in boys before their first birthday. Gene therapy also resumed in the United States with similar restrictions.

The experience with gene therapy in boys with X-linked SCID has created a terrible dilemma for parents, physicians, and authorities charged with the oversight of the research. Imagine yourself a parent of a little boy who has just been diagnosed with SCID. How would you weigh these facts? At this writing, 17 boys with X-linked SCID have undergone gene therapy, 10 at Necker and 7 in the United States and Britain. In all cases, the treatment for SCID has been successful, and the children have regained a functional immune system. Because no one knows how long the transplanted cells and their progeny will work, it is too soon to claim that they are cured of SCID. We must wait several years to see whether their reconstituted immune systems continue to produce functional cells.

Nevertheless, their status—some have been in remission for more than four years—justifies optimism. Until a few years ago, the diagnosis of X-linked SCID combined with the lack of a compatible donor was a death sentence. Today, parents can expect that affected children who undergo the gene therapy will in a few months have a competent immune system. On the other hand, 3 of the 17 (all treated at Necker) have developed leukemia (the most recent is different from the first 2), and 1 of the 3 little boys has died. The French have halted the program at Necker Hospital, and the United States has suspended its three trials. It will only permit gene therapy in boys who have undergone a bone marrow transplant that was unsuccessful. The British have reached a different conclusion about the risks and benefits and are continuing their program.

How does one balance the risk of death from SCID against the risk of death from leukemia? About 20% of boys who are diagnosed with X-linked SCID find a donor who is an appropriate match for bone marrow transplantation (which itself carries a small risk of death). For these children, it seems clear that the proper choice is transplantation. But what of the other 80%? For those who do not have the opportunity to undergo bone marrow transplantation, the risk of death is extremely high. It is impossible for parents not to exist in a state of chronic fear. Long before it kills, the disease takes a great toll.

Should parents of children who do not have a potential bone marrow donor be permitted to consent to gene therapy even though it may carry a 20% risk that the child will later develop a potentially fatal leukemia? Arguments against undergoing gene therapy for SCID include the possibilities that (1) a perfectly matched bone marrow donor will be found, (2) there will be advances in using tissue from imperfectly matched donors, and (3) there will be further advances in gene therapy, especially in regard to regulating the insertion of the gene vectors (thus, reducing the risk of leukemia). The major argument in favor of going forward with gene therapy centers on the daily risk to the child of contracting a fatal infection. But there is another. The 2 children who developed leukemia due to an insertion in the LMO2 gene underwent treatment before the age of four months. Given the age distribution of the 10 children, this is an unlikely event. Because of their knowledge of how the immune system matures, some researchers think that older children are at lower risk of contracting this form of T-cell leukemia. Even in the face of a daily threat of serious infection, the right clinical course may be to wait until the child is at least one year old.

Ethics is not a statistical science. The central principle before each set of parents (the children are far too young to exercise choice) is to decide what is in the best interests of the child. It is hard to make judgments on limited facts, but sometimes we have to. It seems to me that it would be right to forbid gene therapy for children with SCID who are younger than the age that a consensus of experts decides marks a reduction in risk for developing leukemia after gene therapy. With good medical care, most can be kept alive until they reach that age (which is likely to be about one year). I am not comfortable with the decision in the United States to forbid gene therapy to those who had not first tried bone marrow transplantation. Since both therapeutic options are fraught with danger, I think the parents of affected children should have the right to choose.

The saga of gene therapy for X-linked SCID is, despite the setback due to the development of leukemia in two patients, a remarkable success story. First, it was based on solid science and comprehensive animal studies. Second, the physician-researchers acted promptly and appropriately when problems arose. Third, the regulatory bodies responded with appropriate caution but, mindful of the needs of the patients and their families, they worked assiduously to find a means to safely reopen access to treatment. Fourth, scientists probed the mystery of the leukemia and came up with a very plausible explanation and a game plan to circumvent the risk in the future.

The careful investigation of every infant who presents with an apparent genetic disease that is compromising the immune system has led researchers to delineate at least 11 different forms of SCID. In turn, they have so far found mutations in 10 different genes that can cause SCID. This is almost invariably the pattern in our history of delineating several thousand human genetic diseases. The more meticulously we study the patient's clinical condition and his DNA, the more likely we are to describe a new cause for the illness. More research will refine the therapeutic approach. In 2005, researchers reported that gene therapy had failed in the treatment of two older children with a slightly milder form of the disease, suggesting that as-yet-unknown age-dependent factors were important in this approach.

The possibility that gene therapy will become the treatment of choice for a number of similar disorders of the immune system is good. In addition to the successful treatment of most of the boys with SCID, researchers have successfully used gene therapy to treat four children with adenosine deaminase deficiency (ADA). When described by doctors in 1972, ADA became the first of the steadily growing number of immunodeficiency diseases. It is an autosomal recessive disorder that affects equal numbers of boys and girls and accounts for most of the non-SCID patients. Kids with ADA have an illness that is remarkably similar to SCID. Until recently, most died before the age of five from infection. Researchers have been attempting gene therapy in ADA patients for a decade. Follow-up studies have demonstrated that the genetically engineered T cells persist in the patients for at least that long. Over the last few years, improved techniques in the delivery of the engineered cells have resulted in substantial improvement in immune function without serious side effects. Emboldened by success, expert physicians are beginning gene therapy trials for a growing number of immune deficiency diseases. In addition to experiments involving patients with well-known single-gene disorders such as muscular dystrophy and cystic fibrosis, researchers are working on using gene therapy to treat a form of brain cancer and heart disease.

Only about 50–100 boys are born with X-linked SCID in the United States each year. Given the host of far more common health problems faced by children, the triumph over this disorder may not seem great. However, the successful use of gene therapy to treat (perhaps cure) SCID means that research to expand its applications could grow quickly. It is very likely that patients with several less esoteric genetic disorders such as hemophilia, muscular dystrophy, and cystic fibrosis, disorders affecting many thousands of children, will (over the next decade) benefit from this wonderful new field.

PART
3

ANIMALS AND PLANTS

Zoe, a beloved member of the Reilly family.

Dogs

On July 14, 2004, Tasha, a beautiful boxer, made scientific history. She became the first dog in the world to have her DNA sequenced! On that day, a research group called the Canine Genome Mapping Community, of which the unofficial leader is Elaine Ostrander, a geneticist who has spent much of the last decade spearheading the effort to study the dog genome, deposited the first draft of the entire dog genome (about 2,850,000,000 base pairs of DNA) into a freely accessible public database for use by the research community. The success of the dog genome sequencing effort is remarkable in many ways, not the least of which are time and money. It took more than a decade and about $3 billion to sequence the human genome. Building on that work (dogs really are a lot like humans), it took just over a year and $30 million to sequence the dog genome. Although it is essentially the same size as a human genome, scientists sequenced Tasha's DNA in one-tenth the time and at one-hundredth the cost of the human project! With the successful sequencing of Tasha's DNA, the field of comparative genomics is coming of age. With each passing year, we will witness the completion of new sequencing projects at ever lower cost. By comparing gene sequence differences across species, we will gain new insights into gene function and disease predisposition.

Why was Tasha, who hails from upstate New York, chosen? It turns out that, compared with other breeds, the boxer has less variability in its DNA, making it easier to study. A year or two earlier, Craig Venter, who during the 1990s spiced up American science by challenging the federally funded human genome project with a privately funded effort, initiated a dog genome project at a company called Celera. He decided to use the DNA of his pet, a standard poodle. This project produced a draft of the dog genome that probably captured about 65% of the sequence, a major advance, but not adequate to support many projects that researchers wanted to undertake. Venter was quick to provide the DNA sequence data from his standard poodle to the team working on the Tasha project, which considerably eased their task.

Given the many hundreds of interesting and economically important animals and plants that scientists could choose to sequence, how did the domestic dog (*Canis lupus familiaris*) become just the third mammalian genome (after human and mouse) to be sequenced? The answer lies in the intimate history of man and dog. The earliest remains of domestic dogs date from more than 10,000 years ago. However, DNA analysis suggests that the domestication of the gray wolf, the exceedingly close relative of the dog and almost certainly the species from which it evolved, may have begun more than 50,000 years ago. The earliest breeding efforts probably occurred in multiple different human communities across Asia, Europe, and Africa.

At what point over the last 50,000 years the descendants of captive wolves had undergone sufficient inbreeding and selection to become dogs is impossible to answer. However, it may have happened quite quickly. A reasonable guess is that it took at most only a couple of centuries (which is 50–100 dog generations) for the first discernible dog breeds (animals with physical and behavioral traits recognizably different from wolves) to appear. The road to domestication was probably not smooth. As the fortunes of their human masters (who were still living in small groups, eking out a precarious existence) waxed and waned, early dogs may have been abandoned or eaten. The evolution of the domestic dog is further complicated by the fact that there was almost certainly some reproductive contact between dogs and wolves. Some early dogs probably returned to the wild and bred with wolves, thus reshaping the early dog gene pool and giving rise to offspring with descendants who were reintroduced to human society for another round of domestication.

Despite the confidence of some physical anthropologists and molecular biologists that dog domestication started more than 50,000 years ago, today only a handful of breeds can claim a heritage of even 300 years, and very few modern breeds can stake any claim to ancient lineage. Some experts believe there is sufficient evidence that the Maltese existed 2,800 years ago. During the invasion of Britain, Caesar's armies praised the bravery of the mastiffs that fought by their masters' sides. Of the several hundred modern breeds (only 152 of which are registered by the American Kennel Club), the vast majority came into existence within the last 200 years.

The recent and rapid creation of so many distinct breeds (all members of the same species) is precisely why sequencing the dog genome will prove to be of inestimable value to dog and human health. By placing extreme selective pressure on small populations of dogs over hundreds of their generations, humans have been able to create scores of breeds that demonstrate a greater range of physical and temperamental variation than any other mammal. Elaine Ostrander, the prominent dog geneticist, reminds us that a Chihuahua is less than six inches high at the shoulder, whereas the Irish wolfhound measures 36 inches. The Pomeranian weighs 4 pounds, whereas the St. Bernard can easily weigh 150 pounds. The human family has far less variation. Unfortunately, the extreme selective pressure required to select for desirable traits carries a high cost—the spread of recessive disease genes in the closed gene pool.

Domestic dogs are collectively at risk for more than 300 single-gene diseases (such as muscular dystrophy and hemophilia) or highly genetically influenced multifactorial disorders (such as cancer, heart disease, and hip dysplasia). Because the single-gene disorders are usually recessive in nature (i.e., the disease appears in the offspring of healthy adults who each carry one copy of the deleterious allele), it is uncommon that a particular mating can be said in advance to be risky. Because the dog genome is remarkably like the human genome, genetic diseases in dogs are remarkably like their counterparts in humans. Research in dog genetic diseases will teach us a great deal more about the human analogs than ever could be accomplished with similar work in mice. For example, the study of a group of blinding disorders in dogs called progressive retinal atrophies (PRA) has greatly aided our understanding of a group of human diseases called retinitis pigmentosa, including the discovery of a new disease gene.

In their brief to the National Institutes of Health to support the request for funds to sequence the dog genome, Elaine Ostrander, Kerstin Lindblad-Toh, and Eric Lander noted that certain dog breeds are highly prone to develop human-like cancers and that mapping the susceptibility genes (which are almost certainly cognate) could be done much faster in dogs than in humans. They also pointed out that mapping rare traits in dogs is a useful way to better understand related human disorders. Doberman pinschers have a genetic form of narcolepsy that is caused by a mutation in a gene called the hypocretin-2 receptor. Subsequent studies of narcolepsy in humans showed that most cases are associated with a deficiency of the human hypocretin protein.

The dog is also a great model organism to discover quantitative trait loci (QTLs), genes of as-yet-unknown function that by statistical methods can be shown to substantially influence the expression of some prominent physical or behavioral feature. Scientists track down QTLs by crossing dogs with opposite physical features, back-crossing their offspring, and analyzing their DNA to find sequences that track with the physical trait of interest. Given that dog breeds are the product of substantial inbreeding, it may be that just one or two QTLs strongly influence skeletal size (differentiating a greyhound, say, from a pit bull). The fact that such dramatic features as size and body shape may well be a function of the effect of only one or a few genes (which some studies suggest) helps to explain how humans could create so many distinctly different breeds in so short a time.

Historically, dogs have been of incalculable value in the development of safe blood transfusions and in organ transplantation, fields to which clinical research on mice has far less to offer. The dog also has turned out to be a superb organism in which to investigate gene therapy. Recently, researchers used gene therapy to successfully treat seven dogs with a rare glycogen storage disorder. Within their first year, untreated animals with this disorder fail to thrive, lose their ability to stand, develop cataracts, and become afflicted with heart disease. Using a weakened virus to deliver a healthy version of the disease gene to these animals, researchers have been able to keep the animals free of major symptoms of the disorder. Other researchers have delayed the onset of blindness in a dog with the canine version of the human eye disease called Leber's congenital amaurosis. A major dividend from having access to the dog sequence is that comparative studies of genes among three mammals (human, dog, mouse) will greatly enhance our confidence about gene function and gene conservation.

One of my favorite dog genetics stories is based on the work of Gustavo Aguirre, a professor of ophthalmology at the James A. Baker Institute for Animal Health at Cornell's prestigious College of Veterinary Medicine. Since the 1970s, Aguirre and his colleagues, notably Gregory Acland, have made astounding contributions to decoding six separate forms of blindness in dogs caused by retinal degeneration. They have identified which breeds are at high risk, cloned some of the causative genes, developed genetic tests for breeders to avoid at-risk matings, and tirelessly worked to educate community vets and breeders on how to avoid the disorders.

Aguirre's work on a form of retinal degeneration found in Alaskan malamutes has provided profoundly important insight into a rare form of day blindness in humans that is seen in some Irish persons, but mainly in a little-known group of Pacific Island people called the Pingelese. The human disease is called achromatopsia, a term that signifies that afflicted persons have no color vision at all. The Irish call the disease day blindness; the Pingelese call it maskun ("no-see"). Affected people and dogs lack cells called retinal cones. This makes normal daylight seem unbearably bright. Thankfully, the patients can see unusually well at night. They appreciate the crisp beauty of a clear, starry night in a way that those of us with a normal number of retinal cones cannot.

The odd connection between the Irish and the Pingelese is easily explained. Sometime in the 17th century, one or more Irish sailors who carried a copy of the recessive gene fathered children with Pingelese women. The disease would probably have manifested rarely, but for a typhoon that in 1775 nearly wiped out the Pingelese islanders. The natural disaster, which was followed by a famine, reduced a fairly large population to just 20 souls. By chance, several of the survivors carried a copy of the disease-causing gene. Suddenly, a recessive allele for a rare disorder was very common in a tiny gene pool. Today almost 30% of the Pingelese people carry one copy of the disease gene, and about 6% have the disease.

What a typhoon did to the Pingelese, human breeders have done to the Alaskan malamutes. Breeders chose stud dogs that, unbeknownst to them, carried (in addition to many other genes that shaped the most sought after malamute phenotype) the canine version of achromotopsia. The more litters they sired, the faster the disease allele spread in the Alaskan malamute gene pool. By comparing the DNA sequence of the gene responsible for this disorder in dogs and humans, scientists have shown that the Irish and Pingelese have the human version of the dog eye disease.

Thanks to the dog genome project, during the next decade, veterinarians will be able to use scores of new DNA tests to determine whether or not a dog is a carrier of, or affected with (but not yet symptomatic), a large number of single-gene disorders (or at high risk for others). Today, when a knowledgeable dog fancier seeks to buy a purebred puppy, he or she will ask (depending on the breed) if the animal has had an eye exam (for cataracts) or been examined by X-ray to assess risk for hip dysplasia. In the future, one may not buy such a puppy unless it comes with the results of a battery of DNA tests to assure that it is not afflicted with any of a group of genetic disorders.

Canine DNA testing has already taken on great importance at the American Kennel Club. Working with a California genetic testing company, the AKC has developed a DNA identity test which is very similar to that used by the FBI. For a cost of $40, breeders can obtain a test. The results of this will further bolster their claim that a particular animal is the offspring of purebred parents. As of late 2004, the AKC had more than 320,000 DNA profiles in its database. First among the top ten breeds that have had animals "DNA certified" are Labrador retrievers (more than 22,000 dogs tested). The AKC is moving rapidly to use DNA testing in its kennel inspections. Since the test is used to verify parentage claims, it will soon be nearly impossible for a kennel to fraudulently sell a pup alleged to be the offspring of a champion line. Since random DNA testing was added to litter inspections to verify parentage, accuracy of parentage claims has improved from 89% to more than 94%. It is not yet possible to use DNA testing to define breeds, but someday it may be.

Perhaps the most intriguing consequence of the dog genome project is that it will eventually permit the identification of gene variants that have large effects on highly defined behaviors. Given the history of dog breeding, this should not be surprising. Experts assert that most breeds came into being to satisfy a human desire for animals that would reliably fulfill a certain task. They parse the evolution of dog behavior into three major categories: herding behavior (Border collies), guarding behavior (German shepherds), and hunting behavior (spaniels). Many dog fanciers go further, readily labeling breeds with specific traits (which sometimes sounds overly deterministic). As Elaine Ostrander has put it, "Few would contest that dalmatians are high strung, Border collies obsessive, and Doberman pinschers protective. Simple breeding studies have suggested that aggression, herding, and, perhaps, loyalty, are among the canine behavioral traits likely to be controlled, in part, by genetic factors." Ostrander also reminds us that breeding studies have shown that a trait, called "human aversion," which is found in some pointers, is under extremely high genetic influence. The most severely affected pointers can become catatonic in the presence of unfamiliar humans. I anticipate that over the next several decades, dog breeders will focus as much on selecting for desirable behavioral characteristics as they do today on physical traits. Armed with DNA tests, they are likely to be highly successful.

Even though it will take some years to track down major behavioral genes in dogs, we already know enough about the variations in the dog genome that a researcher who is given a sample from a purebred animal (and who has access to the reference database) can tell with 99% accuracy to which of the more than 100 breeds it belongs. Not unexpectedly, the DNA studies that permit this assertion have also challenged deeply held beliefs about dogs. Breeders and show judges define a breed by a collection of physical traits and a pedigree. Dog geneticists look for similarity at the DNA level. Recent DNA studies have led dog geneticists to assert that the German shepherd genome is more similar to the DNA of guarding dogs than to the herding dogs with which it has been traditionally grouped. Within a few years, DNA analysis is likely to be a routine part of certifying that a dog is a purebred animal.

Not so far in the future, it may be possible to purchase a puppy certified not to carry any of the major gene variants associated with an unusual level of aggression. Many of today's breeds (for example, mastiffs, rottweilers, and pit bulls) were bred to be aggressive (e.g., territorial, suspicious of strangers, and quick to attack). According to surveys, about 40% of owners buy dogs primarily for protection. The aforementioned animals (along with German shepherds, malamutes, and huskies) consistently dominate the ranks of the ten breeds most likely to bite humans (an event that occurs more than 4,000,000 times a year in the United States). The behavior of human owners probably has a significant influence on the degree to which an animal expresses his aggressive drive. Yet, it is irrefutable that some breeds carry a much greater risk for seriously aggressive behavior even in the gentlest households. Over a recent 20-year period, pit bulls and rottweilers accounted for one-half of the more than 200 fatal attacks on humans, yet they represent a tiny percentage of all dogs. Genetic testing might someday persuade a family that loves pit bulls to select a less aggressive animal.

An experiment that has been going on for 50 years in Novosibirsk, Russia, is a harbinger of things to come in the field of behavioral genomics. In the 1950s, a scientist named Dmitry Belyaev set up a colony of silver foxes with the intent of partially domesticating them so that furriers would have an easier time working with them. He chose his founding population on the basis of degree of tameness. Although that was the only phenotype selected for, in subsequent generations, the animals began developing wider skulls, shorter snouts, curly tails, floppy ears, and new coat colors. That is, they began to look more like domestic dogs. The scientific team developed an elaborate method to score the degree of tameness versus wildness in the animals. By the tenth generation, about one-fifth of the animals in the colony were almost as tame as domestic dogs; after 35 generations, about 80% of the animals were in that category. The scientists are now using genomic analysis to identify the genes that drove this rapid change in phenotype.

Given the status that dogs hold in so many millions of families, it is hardly surprising that an entrepreneur is trying to commercialize canine genomics by attracting pet owners. A California company called K9Genetics Corporation is now selling personalized dog chow and personalized dog treats with the claim that it can use DNA analysis to pick the most nutritious diet for beloved pets. The owner need only fill out a brief questionnaire, swab some cells from the dogs cheek, and send them off for analysis. The company has not yet disclosed the genes that it analyzes or the research upon which it bases its claims.

Of all model organisms, it is likely that studies in dogs will most legitimize the nascent field of human behavioral genetics. During the next decade, watch for claims reported in the media that a research group has cloned the genes for shyness, aggression, herding instinct, and several other complex behaviors (retrieving) in dogs. Of course, the expression of such traits is also dependent on and shaped by environmental factors, but the genetic influence is likely to be very high. (My then 8-week-old Labrador puppy, Cassie, started retrieving items the first time we threw them!) The world of comparative genomics will soon be so sophisticated that animal researchers will be able to find the human counterpart of the dog behavioral genes simply by doing a Web search. These discoveries will almost certainly catalyze another round of impassioned debate about the grand questions of determinism and free will. They may also provoke anxious demands that scientists not be permitted to probe humans to discern genetic drivers of behavior. This will be a tempest in a teapot. Our brains are so sophisticated and our behavioral repertoires so complex that no sensible person will map knowledge about dog genes directly to human behavior.

Studies of the genetics of dog behavior inexorably pull us to reflect on the notion of dog intelligence, an admittedly amorphous and ill-defined concept. Dogs have long been bred for certain qualities that humans associate with intelligence. During World War II, the United States Army conducted research on which breeds could be most useful at certain tasks. Of hundreds of breeds, they "drafted" only five: German shepherd, Belgian sheepdog, Doberman pinscher, collie, and schnauzer. During the 1950s, researchers at the Jackson Laboratory in Maine conducted extensive training experiments on the capacity of breeds to perform defined tasks. They found that fox terriers did best on tests that called for confidence in strange situations, cocker spaniels did best on obedience training, and African basenjis were best on tests that required independent action. One measure of canine IQ might be the number of commands that an animal can demonstrably obey. Of course, that suggests that in some sense the dog is able to understand a spoken word. Is that really possible? Perhaps Rico's story will convince you it is.

The Smartest Dog in the World

A lot of devoted dog owners may take umbrage at this news. Your clever pet—the fox terrier that can catch a Frisbee 10 times in a row, the overprotective collie that will not let your children venture near the road, the wise old retriever that senses your arrival a full minute before the family car turns the corner—is not even remotely as intelligent as Rico, a Border collie that has easy familiarity with a vocabulary of more than 200 words. The Border collie originated in Northumberland on the lowlands that mark the boundary between Scotland and England. The breed descends from other droving breeds that were used by shepherds. Although bred primarily for agility and speed (traits essential to keeping 50 sheep in line), Border collies have long been more generally appreciated as highly trainable dogs that respond well to voice commands. Breeders often describe them as perfectionists who hate making mistakes.

Rico is, as far as I can tell, the only dog that has been the subject of a scientific report in *Science*, one of the two or three top scientific journals in the world. The article reports on an experiment to determine whether

Rico is able to acquire a rough understanding of a new word after only a single exposure to it, a process known as "fast mapping" that is a quintessential aspect of the means by which children rapidly expand their vocabulary. Equipped with this skill, English-speaking two-year-olds learn about 10 new words a day, building to a vocabulary of 60,000 words or so by age 18. Rico, who took about ten years to learn 200 words, has not done nearly so well as humans. Nevertheless, his understanding rivals the best performances of apes, dolphins, and parrots.

Just as a child with great genetic potential benefits from a nurturing environment that permits his talents to blossom, Rico was blessed with unusually patient owners who began working to develop his skills when he was a few months old. They report that the key tool in developing his vocabulary was the game of "fetch." They began with just a couple of items that they named several times over. They then hid them in the room and commanded Rico to find. They regularly added items, simply naming them a few times and then permitting Rico to play with them.

The scientists were fascinated by reports of Rico's prowess because they perceived that testing him might help resolve a long-running debate among language experts over whether fast mapping was mediated by general learning and memory mechanisms (which would lead one to expect animals other than humans to demonstrate this trait) or was a device specific to language acquisition (which would mean that no animal should demonstrate it). Essentially, what they were most eager to explore was the speed with which Rico could learn new words.

The first step was to confirm that Rico really had the immense vocabulary that his owners claimed. The scientists randomly divided the 200 items that he could recognize by name into 10 sets of 20. They then hid a set of items in an experimental room while the owners and dog waited nearby. The scientists then joined the owners and instructed them to command Rico to fetch 2 items among the hidden objects. While he was searching for the requested objects, Rico could not see his masters or the scientists. Working with 2 sets of items, Rico correctly retrieved the commanded item 37 out of 40 times. The odds of this happening by chance are less than 1 in 1,000. To those language theorists who might assert that Rico interpreted all commands as single words (such as "fetch the sock"), the scientists pointed out that Rico could follow more complex demands, such as "bring the ball to David" or "put the shoe in the box."

The results of the fast mapping experiments are the most interesting. In these exercises, the researchers placed a novel item into a set of 7 known items and, after briefly introducing the new object to Rico, hid the set of 8 in an adjacent room. They used a total of 8 novel items in 8 sets of objects and conducted 10 "identification" sessions. The owner was told to first ask Rico to fetch a familiar item and then command him to fetch the novel item. Rico fetched the novel item on his first effort. Throughout the entire series of sessions he picked correctly in 7 out of 10. Here, too, the chance that this was the result of a random selection by Rico was less than 1 in 1,000. Rico knew enough to figure out that the novel item could be thought of as "not any of the ones I have known for a while."

Interested in knowing how well Rico remembered novel items which he had seen on the day of the experiment, the scientists retested Rico four weeks later. During that period, he had not seen the novel items. This time, the researcher placed the "target item" among a set of 9 others that included 4 long-familiar items and 4 novel items. After first asking him to fetch a long-familiar object, they next asked him to fetch the object that he had seen once four weeks earlier. He did so correctly 3 out of 6 times, a success rate that would occur by chance less than 1 in 10 times. When retested a few minutes later, his success rate rose to 4 out of 6, a score that would occur by chance less than 1 in 50 times. Those who are unimpressed with Rico's performance should note that it is equivalent to the abilities of a three-year-old human.

The researchers, a trio led by Juliane Kaminski at the Max-Planck Institute in Leipzig, Germany, concluded that the Rico experiment proved that the ability to attach meaning to sounds evolved long before the ability to produce specific sound patterns as language. Put another way, some of the cognitive abilities necessary for speech were in place in the brains of animals long before there were humans.

Of course, however impressive Rico's abilities, they fall far short of the vocabulary of any normal child. Children learn words far more easily than does Rico. For example, humans can learn words merely by overhearing a conversation in context; Rico only learns by playing fetch with his owners. Could it be that Rico is learning words in somewhat the same fashion as a toddler, but not as well? Or, as many would argue, does Rico learn in a far more restricted way than does a child? Children, after all, can understand a word in many contexts once they have learned it, whereas Rico seems only able to learn a word as a fetchable item. Paul Bloom, professor of psychology at Yale who wrote an editorial about Rico for *Science*, points out that Rico's skills would be far more impressive if he could follow a command to do other than fetch a small object.

I see little to be gained from debating how closely Rico's capacities approach those of humans. I choose to reflect on Rico's abilities from a different perspective. As the scientists put it, are Rico's accomplishments "based on an exceptional mind" or merely the result of intensive training in assessing word-object combinations? Is Rico a canine genius, or can any healthy Border collie be trained to match his performance?

The question of whether Rico is a canine genius is in part an empirical question that could be investigated with relative ease. One could search the world for Border collies (or other dogs) about which their owners make equivalent claims and then test a cohort of them using the same experiments. If one cannot find a dog that possesses even one-half of Rico's fetch vocabulary despite roughly equivalent training, one might be justified in concluding that Rico is a genius among Border collies.

Perhaps Rico will become a highly desirable stud, the patriarch of a line specifically bred for fetching skills. Imagine that Rico mates with a series of females that are also accomplished "fetch" linguists. Among their progeny might we find sons or daughters that with similar training might easily eclipse Rico's records? Could such a breeding program redefine the domain of duties that dogs perform on behalf of humans? Imagine a seeing-eye dog that could respond correctly to several hundred simple commands! Over the next decades, humans may be able to develop dog breeds that eclipse current levels of intelligent behavior.

SNUPPY

I would be remiss if I did not add a few words in response to the news that a team of researchers at Seoul National University in South Korea has recently reported that it has cloned a dog. Dr. Woo Suk Hwang, the veterinarian scientist who spearheaded the effort, named the pup, an Afghan hound, Snuppy, which is an acronym honoring the university where the research was done. Given the peculiar nature of canine reproductive biology, this is an incredible feat, one that other scientists have called, "the Mount Everest of cloning." To clone a dog, one needs access to fresh dog eggs, from each of which one removes the haploid nucleus and replaces it with a diploid (full set of genes) nucleus from a body cell of the adult animal to which the clone will be an identical twin. But dogs only ovulate about twice a year, and there is no reliable method to anticipate when ovulation occurs. In their breakthrough paper, the South Korean scientists claim to have discovered a biochemical marker that signals when a dog is ovulating. When dogs ovulate, the eggs are immature, difficult to find, and hard to manipulate. The process of cloning animals usually includes a step during which the newly created embryo is grown in culture before it is placed in the uterus of a surrogate mother. However, thus far, no one has been able to culture early dog embryos. Dr. Hwang and his team reported that they overcame this key problem by performing surgery to place the eggs containing the diploid DNA into the surrogate dogs almost immediately after completing the DNA transfer. They claimed that they had an operative window of only four hours before the cloned embryos began to die. The South Korean team reported that they harvested 1,095 eggs which they then surgically transferred into 122 surrogate mothers before they finally were rewarded with an apparently healthy cloned puppy. Only two other pregnancies were achieved; one was a miscarriage and the other resulted in a puppy that died shortly after birth.

Unfortunately, late in 2005, Dr. Hwang confessed to submitting fraudulent reports of his work with stem cells, a disclosure that shook the research community. This called into question whether the work that resulted in the birth of Snuppy was true. The university conducted a scientific inquiry which concluded that Hwang really had succeeded in cloning a dog. Even with initial success, canine cloning is still a long way from commercial use. Scientists estimate it would cost about $1 million to repeat Hwang's work that led to Snuppy. Still, it is likely that with time the cloning process will become more efficient. Someday, man's best friend will make crucial contributions to understanding the many diseases that burden both species.

Little Nicky, a cloned cat. (Courtesy of Genetic Savings & Clone.)

Cats

My wife, Nancy, and our children share our home with three cats. George is an elderly calico who weighs next to nothing, is hearing impaired, and sleeps 18 hours a day; Sapphire, a beautiful ebony with fur like the finest silk, loves to have her belly rubbed and is seriously overweight; Mooka, only two, is a lithe black and white that dependably deposits dead voles on our doorstep just before dawn. Over the last 20 years, there have been a number of others. Sadly, especially for our younger son, two years ago we lost a kitten to feline leukemia virus. A few months ago, my favorite cat, a hard-living orange male named Bailey, who really did believe he had nine lives, was killed by an automobile. I am quite sure it was not the driver's fault. One measure of Bailey's extreme arrogance was his fondness for sleeping in the road.

Although I have lost count, I know that from his birth, Bailey regularly taunted and (except for the last episode) cheated the grim reaper. He came into our lives during a summer vacation in Maine 10 years ago. We were leaving Bailey's Island when we saw a sign advertising free kittens propped up by a tree in front of a shabby trailer. When the woman at the door (who was no fool) saw the kids and announced that if she did not find a home for the kittens that day she would drown them, my children made it clear that one kitten was returning to the car with us. During his 10 years, Bailey was by far our most expensive cat. When he was about 3 he was struck by a car and flung over a five-foot hedge some 30 feet into our yard. After extensive orthopedic surgery on his back legs, he made a full recovery. An almost identical accident with quite similar surgical fees took place a couple of years later. The summer before he turned 7, Bailey disappeared.

We searched the neighborhood, posted the usual plaintive signs on telephone poles, and kept our spirits high. Hope waned. Two weeks had passed when a couple who lived down the street and who had been away for that period showed up at our door with an emaciated Bailey. He had apparently sneaked into their cellar just before they left for vacation and had been incarcerated without food for two weeks. As far as they knew, he had not even had access to water. Perhaps he survived on mouse blood. My favorite memory of Bailey is that late at night when every other human and animal in the house was asleep, he would find me reading or in front of the television and insinuate his well-muscled body and gorgeous orange coat onto my chest, stare directly into my eyes, and purr contentedly. He never did this when anyone else was about, as if it would erode his carefully honed macho image.

Those who are not cat lovers will at the least shake their heads over the college tuition payments I diverted to Bailey's health care. I am, however, unashamed. Why? I take comfort in the fact that in the United States alone more than 60 million cats have the run of human dwellings. The situation in Europe is not much dissimilar. Several of the thousands of Web sites devoted to them estimate that the earth is home to 500 million pet cats, far eclipsing the number of dogs. Again, why? What is it about cats that reduces so many humans to such a servile condition? They are almost uniformly beautiful creatures. They have gorgeous hair, entrancing eyes, (in most cases) lithe figures, and strong personalities. They have an innate sense of how to balance aloofness with companionability. Everyone agrees that kittens are irresistible. The success that cats have achieved in dominating humans is a fine example of coevolution.

Humans have been so thoroughly brainwashed by cats that when asked about feline origins, the vast majority of "owners" proudly note that the cat first became a human pet in Egypt some 5,000 years ago. In fact, their history with humans is much deeper in time. In April 2004, a group of French archaeologists reported in *Science* its discovery in Cyprus of a cat (*Felis silvestris* cf. *lybica*) skeleton buried just 15 inches from a grave that could (by radiocarbon dating of the nearby human remains) be accurately estimated as being 9,500 years old. Replete with polished stones, axes, ochre and flint tools, the human grave appears to have been the final resting place of a prominent person. Meticulous examination of the site indicated that a separate grave had been prepared for the cat. Because its skeleton was found intact and in the same stratigraphic layer, there is a strong reason to infer that the animal was sacrificed to accompany its master in the afterlife. The logical conclusion is that, in Cyprus nearly 10,000 years ago, cats were not merely used to guard granaries against murine invaders. They had already cleverly adapted to become household pets, appreciated for their beauty and personality and, doubtless, not required to do much work. This has to be one of evolution's great success stories!

We know for certain that domesticated cats did not arise in Cyprus, for it is an island that lies about 50 miles from Turkey, the nearest landmass inhabited by wild cats. Given the ubiquity of several small feline species on the Asian continent, it is likely that the domestication of cats unfolded in several areas as humans developed agriculture. Indeed, there are a few Neolithic cave paintings depicting cats that date to the earliest period (10,000–12,000 years ago) in which we can confidently accept that humans were engaged in farming. Simply put, grain attracted rodents and rodents attracted wild cats. It is likely that humans quickly grasped the benefits of cats in warding off rodents. Cats may have been equally astute in recognizing the comforts of human society.

Once the connection was established, the cat became part of the human diaspora. They were popular in Egypt more than 5,000 years ago, they were commonly found among humans throughout China nearly 4,000 years ago, and they were treasured in Islamic societies (in which legend holds that Mohammed adored them) more than 1,000 years ago. Cats may have reached an apotheosis in Egypt. A religious order centered on cat worship flourished for 2,000 years. The goddess, Bastet, had the body of a woman and the head of a cat. Some sects worshipped cats in Roman times, and the remains of cats have been found in human habitation sites in the Ukraine that date to the 7th century. In 10th-century Britain it was a crime to kill a cat. In the 12th century, a royal decree required that British farmers own and provide for at least one cat, so indispensable was the feline thought to be to protecting grain.

Domestic cats are not mentioned in the Bible (but there are nearly 100 references to lions). Although the historical record is sparse, there is some evidence that until the late Middle Ages, cats lived in harmony with Christians. Church records indicate that for a time they were the only kind of pet allowed in convents. However, during the Early Middle Ages, some pagan tribes began to deify cats. This may have been a factor in the subsequent troubles that cats had with Christian authorities.

The social compact between humans and cats, which had gone remarkably well for several millennia, began to decline in France in about the 12th century with the ascendance of folklore that cats were the agents of the devil. It reached its nadir during the emergence of cities in early modern Europe, especially southern Europe. For nearly three centuries, the lower orders of urban society regularly taunted, maimed, and tortured cats in great numbers. Cat torture and killing was even a relatively common theme in literature from the early 17th to the 19th century. Cat killing was so common and so widespread in 18th-century Europe that some historians have characterized it as a leisure activity of the poor.

The prominent French cultural historian, Robert Darnton, wrote a book, *The Great Cat Massacre*, based in part on the memoirs of one Contat, an apprentice in a Parisian print shop in the 1730s. Contat recounted that the most hilarious event of his tenure was the day his colleagues ran amok among the neighborhood cats. What explains widespread behavior that we would now universally regard as cruel and criminal? Darnton is certain that in the 18th century cat killing was not regarded as sadistic. He notes that, in addition to being feared as agents of the occult, city cats were of considerably less economic value to humans than were their agrarian cousins. Howling urban strays disturbed everyone's sleep, and the poor almost certainly deeply resented that pets of the rich ate better than they did. Depending on one's point of view, cats were either in league with the devil, or vermin, or symbols of the ruling class. But even in the days of feline persecution, many well-to-do people kept and protected cats. The great lexicographer, Samuel Johnson, was a cat fancier, as was George Washington. Cat killing ceased to be an acceptable pastime in Europe and America during the late 19th century.

Taxonomists have been arguing about the evolutionary history and classification of cats since the days of Linnaeus. He originally classified all cats—from lions to leopards to lynx to house cats—as members of a single genus, *Felis*. Over the next two centuries, his academic descendants repeatedly revised the scheme, reaching a zenith of 23 genera, including a total of 36 species. In 1996, the Felid Taxonomy Advisory Group radically regrouped the world's cats (Felidae) into three subfamilies: Felinae includes 13 genera of small cats; Pantherinae comprises 4 genera of large cats; the cheetah is the only member of the Acinonyxchinae subfamily. Today scientists recognize 36 species of wildcats. Debate rages about the number and definition of subspecies thereof (some taxonomists recognize 20!). Doubtless, the 1996 classification will also be revised, for it was compiled before scientists had access to studies comparing cat DNA sequences. For example, we now know that the black panther is really a puma with a variant in a single gene that controls coat color.

Studies of the fossil record, combined with comparative studies of the chromosomes of mink, marten, badgers, and pandas, suggest that the "universal" carnivore ancestor arose about 75 million years ago. In the intervening millennia, several genera of cats doubtless rose and died off before the 18 modern genera became established. There is a rough consensus that domesticated cats are descendants of one or more subspecies of *Felis silvestris*, the forest or wildcat that still inhabits parts of Europe and Asia. *Felis silvestris* gave rise to three lineages: the African wildcat, the European wildcat, and the Steppe wildcat. The cat family (Felidae) demonstrates an astonishing range in size. The smallest is the rusty-spotted cat found in Sri Lanka, and the largest is the Siberian tiger that weighs about 500 times more than its little relative.

In sharp contrast to the dog (which has 78), domestic cats have 38 chromosomes. In recent years, we have come to understand that differences between species in chromosome number does not permit one to infer much about the magnitude of genetic divergence at the level of the DNA molecule. Long stretches of a particular chromosome in one species often carry gene sequences that closely match a stretch of genes found on a different chromosome in some other species.

How did the domestic cat arise? The story is almost certainly quite similar to the history of dogs. Over the last 10,000 years, humans repeatedly captured feral cats and consciously or unconsciously selected among their offspring for desirable traits. Surely, there were innumerable instances of contact between early iterations of the domestic cat with its wild cousins, a process that permitted repeated rounds of selection from among animals with robust genetic diversity. Even though they wanted animals to guard the grain, early cat breeders almost certainly applied selective pressure in favor of less aggressive animals. Unusual expressions of coat color and texture, eye color, and body shape, the physical traits most prized by cat fanciers today, probably intrigued humans from the outset.

Of the approximately 3,000 varieties of house cat in the world today, only about 240 are recognized as breeds with pedigrees. The Cat Fancier's Association, which is the world's largest registry of pedigreed cats, currently recognizes just 37 pedigreed breeds for showing in its championship class. Although its origin is mysterious, no breed is of greater historical importance than the Siamese. Victorian England fell in love with cats, and by the 1880s, breeding and entering prize animals in competitive events was the rage. The British discovered the Siamese in Asia in the late 19th century, and it was a competitor in the Crystal Palace Cat Shows in London by 1881. Over the next 80 years, intense breeding efforts led to four classes (seal point, blue point, chocolate point, and lilac point) being recognized as competitive categories. These color variations are due to the effects of a temperature-sensitive enzyme that restricts the development of color to the coolest regions of the body! During the 1950s and 1960s, Siamese cats were the most popular breeds in many countries. Of course, most of us share our home with common house cats. They may not be as elegant, but they have excellent genetic diversity!

CAT GENOMICS

Feline lovers will not be happy to learn that until recently scientists had devoted more attention to understanding the dog genome than they had to the cat genome. Fortunately, those interested in cat genetics have a champion—Stephen J. O'Brien, the Chief of the Laboratory of Genomic Diversity at the National Cancer Institute in Bethesda, Maryland. O'Brien, who was the leading proponent of a Cat Genome Project for many years, offered five reasons why we should spend the millions of dollars that would be needed to sequence *Felis catus*. The first and most obvious is that humans keep about 500 million cats as pets. As domestic breeds of cat are, like dogs, the result of intense selective pressure, cats are burdened with many (more than 200 have been described and the list grows monthly) recessive genetic disorders. A full understanding of the cat genome coupled with study of protein function would result in the rapid development of many new genetic risk tests and many new leads for better therapies. Given the billions of dollars spent on care of cats, knowledge that improves their health and wellness would certainly confer economic benefit.

The second reason is that, like the dog, the cat is an excellent model organism in which to study human hereditary diseases. We are sufficiently close on the evolutionary tree that many single-gene disorders which afflict cats also afflict humans. A short generation time, the ability to control mating, a relatively large size, and a similar physiology make cats good surrogates for clinical study. Third, cats are, unfortunately, burdened with two viral infections that cause diseases quite similar to human leukemia and AIDS. Study of the cat immune system could give important insights in the treatment of the human disorders. For example, cheetahs become infected with, but are not made ill by, the feline leukemic virus that kills house cats with an AIDS-like syndrome. Fourth, the organization of the cat genome is comparatively close to that of human, permitting studies of the evolution of the mammalian genome. Fifth, the genetic traits that attract fanciers to any one of the 37 breeds that are recognized by the Cat Fancier's Association are both highly polymorphic and (in some cases) linked to disease genes. Understanding these genes at the sequence level could greatly assist breeders in producing more desirable and healthier champions. Finally (and to my mind, most significantly), many of the world's 36 species of wildcat are endangered or threatened. Full knowledge of a cat genome could greatly aid in protecting them.

In 2005, O'Brien and colleagues completed what is called a "third generation" map of the domestic cat. They used a variety of techniques to localize more than 700 short tandem repeats (STRs), short stretches of DNA that mark 20 or more spots on each cat chromosome. They used these markers to conduct linkage studies, allowing them to identify the location of more than 1000 genes. In addition, they were able to use STRs to investigate whether it could be shown that some breeds are considerably older than others. So far, they have not been able to do so. However, they have been able to study the distribution of markers among breeds and to develop algorithms that allow them to correctly assign a specific cat to the correct breed in more than 90% of cases.

In 2004, the National Human Genome Research Institute chose the cat as one of the species—along with the elephant, the orangutan, the shrew, the hedgehog, the guinea pig, the tenrec (a small mammalian insectivore from Madagascar), the armadillo, and the rabbit—for which it would pay to have genomes finely mapped and fully sequenced. Agencourt Bioscience Corporation, a Massachusetts company, organized a team of about 100 scientists and technicians to do the work. Cinnamon, a very attractive Abyssinian that is a resident of a long-studied colony of inbred house cats maintained at the University of Missouri, had the honor of providing blood for the project. This made sense. Her lineage can be traced for decades (back to Sweden), and much is known about the health and illnesses in her ancestry.

In late 2007, the scientific team announced that it had completed its effort to construct a "light" sequence of Cinnamon's genome. Because of its complexity, it is usually necessary to sequence a genome 6–7 times to arrive at a consensus sequence that is definitive. A light (two-fold) sequencing effort yields most of the sought-for information, but it is by no means a perfect read. However, because of the great similarities among mammalian genomes, it was possible to draw on knowledge generated from heavier sequencing of human, chimp, mouse, rat, dog, and cow to refine the analysis of the cat genome. The current best estimate is that a cat has 20,285 genes, almost precisely the same number as do you.

One example of a genetic disorder that burdens cats and humans is hypertrophic cardiomyopathy (HCM), a single-gene, dominantly inherited disorder that is particularly common among the Maine coon and the American shorthair breeds. HCM is diagnosed in cats by clinical examination and ultrasound study of the heart. In affected animals the heart muscle is unusually thick, compromising the flow of blood out of the heart when it pumps and constricting the coronary vessels that deliver blood to the heart muscle. A chest X-ray and EKG study provide valuable diagnostic clues. For breeds in which the disorder is common, veterinarians advise screening animals for signs of the disorder before they are allowed to mate and removing affected individuals from the breeding stock. Researchers have not yet demonstrated that the mutations which cause HCM in cats also cause it in comparable human genes, but it is likely that this is the case. There will soon be a DNA-based test for HCM in cats. If it is widely and carefully used, and if affected animals are not allowed to reproduce, the frequency of the disease could be nearly eliminated in just a few decades.

The sequencing of the cat genome is likely to aid the growth of gene therapy. Like humans and dogs, cats suffer from a variety of diseases called storage disorders. In each of them, a mutated gene programs a defective enzyme that cannot break down some important carbohydrate or fat. This causes cells to swell and become dysfunctional. Those interested in gene therapy are especially drawn to storage disorders, for if one could deliver a normal version of the gene into cells and if the newly incorporated gene produced its protein, it could break down the accumulated molecules and, possibly, reverse some aspects of the disease. One enterprising group of veterinarians has used gene therapy to treat feline mannosidosis. They attached the normal version of the gene to a benign virus and injected it into the brains of affected cats. They then observed them for several months, comparing them to untreated cats with the same disorder. The treated cats showed remarkable improvement, and MRI studies of their brains showed that their brain lesions were shrinking!

Although scientists know less about the genome organization of cats than of dogs, they have had more success with felines than with canines in the controversial science of cloning. Oddly enough, the first successful cloning of a cat, which was achieved by a group at Texas A&M University in December of 2001, is a direct result of a project that was initiated to clone a dog. In 1997, John Sperling, a former college professor who in two decades grew his idea for an online university into the University of Phoenix, a multibillion dollar, publicly traded company, was inspired by the cloning of the sheep, Dolly, to fund a similar project to clone his beloved dog, Missy (nicknamed Missyplicity). Now deceased, Missy was three-quarters Border collie and one-quarter Siberian husky.

To oversee the cloning project, Sperling created a California (where else?) company named Genetic Savings and Clone, Inc. Although its goal was to establish the world's first pet cloning service, during its first five years GSC, aware that the technology was not yet mature, was content to fund the research and offer tissue banking services so that beloved pets might be cloned long after they died! The arrival of the first cloned cat (named Cc), a blue-eyed tabby with a white belly, brought the commercial vision a big step closer to realization. Not surprisingly, it also generated a barrage of criticism. Some critics focused on the huge numbers of abandoned cats that were euthanized each year, arguing that the $50,000 cost of cloning would be much better spent on cat welfare programs. Others argued that it is unethical to force cats to serve as surrogate mothers (to carry the cloned animals to term). Most importantly, knowledgeable scientists argued that currently available cloning methods carried a high risk that many of the liveborn clones would live with impaired health.

GSC persevered, awarding more than $4 million in grants to the scientific team at Texas A&M to continue its work in pet cloning. In June of 2004, two more cloned kittens were born, each to a different surrogate mother. The kittens, Tahini and Bab Ganoush, are genetic duplicates of Tahini, a female Bengal cat belonging to the son of the company CEO, Lou Hawthorne. Cat clones numbers two and three are the product of a new technology called chromatin transfer that treats the donor cell to help it dedifferentiate to a primordial-like stem cell. The kittens were sufficiently healthy at four months to make a celebrity appearance at New York's annual cat show.

In December 2004, GSC announced that it had produced its first commercially cloned cat. In the flick of a whisker, California animal rights activists inundated legislators demanding that a law be enacted to forbid the sale of cloned pets. They argued that because cloning can lead to the creation of animals with degenerative disorders, it constitutes cruelty to animals. A bill was filed to end the practice, but it did not receive much attention. If enacted, it would join the existing California law forbidding the sale of "GloFish," transgenic aquarium fish which, thanks to a transgene that codes for a fluorescent protein, glow in the dark.

In 2005, GSC opened a state of the art laboratory in Madison, Wisconsin, dropped its price for cat cloning to $32,000, and launched a major marketing campaign. It also claimed to be close to success in its effort to clone dogs, and announced that it would offer tissue banking for those pet owners who were worried that their favorite dog would die before cloning became possible. Whether because of the high price tag, ethical objections or aesthetic revulsion, only a handful of people ever used the services offered by GSC. Late in 2006, it closed its doors. I have no doubt, however, that a few years from now a new company (dare I say a clone?) will re-enter the market with cloning priced in the range of $1000, and that there will be many customers.

Far more important than the use of genetic science to clone dead pets is its application to help save endangered species. Among Steve O'Brien's many scientific contributions to humanity, the one for which I am most thankful is his work on the conservation of the big cats. From Asiatic lions to the Florida panther, many species and subspecies of the big cats face extinction. Dr. O'Brien and his team have been instrumental in using DNA studies to show that in some subspecies genetic diversity is so low (due to a tiny breeding population) that the danger of succumbing to recessive genetic diseases is high. Their research findings can be used to develop conservation strategies that just might prevent extinction.

The Far Eastern or Amur leopard (*Panther pardus orientalis*) is reduced in the wild to a relict population of fewer than 40 individuals in the Russian Far East. O'Brien's team analyzed mitochondrial DNA to substantiate the distinctiveness of this subspecies, but in so doing also documented the lack of genetic diversity. Fortunately, there is at least one captive population of Amur leopards that has significantly greater genetic diversity and could be used as a source of mates to enhance the genetic health of the wild population. However, the captive population is derived from a genetic admixture of the Amur leopard and a Chinese subspecies (*Panthera pardus japo-*

nensis). Thus, any effort to save the wild Amur leopards from extinction will counter criticism from purists that the use of a significant number of Chinese parents will erode or perhaps destroy the integrity of the threatened subspecies. I think this is the necessary evil that we must endure. Blurring of contested boundaries around a threatened big cat is preferable to not having the cat exist in the wild at all.

A convincing example of the successful use of a closely related subspecies to help save an endangered group of cats (in whom one-third of the deaths are from being struck by motor vehicles) is the introduction of female panthers from a Texas population into the breeding territory of the Florida panthers. By the 1990s, the genetic diversity of the Florida panther was dangerously low. For example, nearly half of the males were born with undescended testicles, probably due to a recessive disease. The Texas females were introduced in 1995, and none of their Florida offspring have been born with this problem.

Some species of big cats in Asia face far greater danger than inattentive drivers in Florida wilderness areas. In India, which is still home to wild populations of lions, tigers, and leopards, the major threats are loss of habitat and illegal poaching to obtain body parts that are prized as traditional Asian medicines. There is not much that scientists can do about disappearing habitat. But molecular biologists in Hyderabad, India, are using tools developed in human forensics that might someday aid in the prosecution of those who illegally traffic body parts of the big cats. They can use DNA markers to show that the remains of a particular animal came from a species that it is illegal to hunt or even from a group of animals that is found only within a protected game park.

Of the 37 species of cats in the world today, only one—the domestic cat—is not endangered. While environmental groups work to preserve habitat, molecular biologists are working to understand how to maximize the chance that dwindling populations can be managed to avoid genetic diseases that arise from inbreeding and to maximize diversity without violating established boundaries that define varieties. It is not too late to use the results from Cinnamon's sequence to help save her wild relatives from extinction.

A laboratory mouse getting his exercise by climbing a model of DNA. (Reprinted, with permission, from Nagy et al. 2003.)

Mice

Despite our perennial sense of pre-eminence in the biological world, humans are markedly similar to rodents.

Joseph Costello, *Nature Genetics* **37**: 211, 2005

It was a crisp blue day in Bar Harbor—the kind you get the day after a thunderstorm clears the humidity. Underneath the giant white tent, a small army of workers was busily setting up tables and consulting charts. Others were unloading boxes from a truck. A wedding reception? Good guess, but wrong. The staff at the Jackson Laboratory was preparing to show off their mutant mice to a hundred or so geneticists who were attending the annual short course on mammalian genetics. For an hour or so, I wandered from table to table, examining dwarf mice, deaf mice, mice with coat color mutations (including one that is butterscotch yellow), obese mice (the causative gene is called "tubby"), and mice with a variety of neurological disorders. One of the newest mutations to be characterized is a mouse version of Charcot-Marie-Tooth disease. At about five weeks of life, the affected animals start to lose strength in the distal muscles of their hind legs. Like so many mouse genetic disorders, the similarity to the human disease is remarkable. The scientists at the laboratory work assiduously to describe the newly discovered mouse disorders and then place the descriptions on the lab's Web site so that researchers the world over can make use of them.

From conception to birth, a mouse develops in just 20 days. An adult mouse weighs about 25 grams; six mice weigh about the same as a baseball. Put another way, if the mouse were a unit of measurement, a moderately large man would weigh 4000 mice! The mouse heart beats about 630 times a minute, almost 10 times faster than the human heart. Incredibly, mice take about 160 breaths a minute (more if a cat is nearby). On the other hand, mouse systolic blood pressure (the higher number of the two) is similar to ours—in the range of 80–160 millimeters of mercury—and mouse body temperature is the same as human. Mice drink only about a teaspoon of water a day, and they pee out only a third of that. A mouse is sexually mature at four weeks. Females typically have six litters a year, with each litter having anywhere from one to ten pups. In the laboratory, mice (at least those that are not sacrificed in experiments) live up to three years; in the wild most do not do so well.

The mammals arose sometime between 80 and 100 million years ago (fossil evidence supports the lower number, but DNA evidence suggests the higher one). For perhaps 10 million years, our early mammal ancestors (mostly creatures that look vaguely like today's small insectivores) lived in the nooks and crannies of a world dominated by the dinosaurs. But, about 65 million years ago, coincident with a dramatic die-off of the dinosaurs, likely caused by a catastrophic meteor strike (which made the earth inhospitable for many years), the mammals became ascendant. On the basis of fossil studies, evolutionary biologists tell us that since their origin mammals have undergone relatively rapid rates of speciation. Today, about 4,600 mammalian species share the planet with their human cousins. Together, we constitute the current iteration of the mammal success story.

On the evolutionary tree, the mammalian limb that gave rise to the common house mouse (*Mus musculus*), the subject of this chapter, has many branches. Today, the mice and rats of the world are categorized in two distinct families; the Cricetidae comprise the New-World species and the Muridae comprise the Old-World species. There are more species within the Cricetidae than in any other family of mammals.

The line of mammals from which modern mice descended and the line that gave rise to humans diverged from a common ancestor about 60–70 million years ago. Even on the evolutionary scale, that is a long time—enough to permit the emergence of significant differences between humans and our murine cousins. For example, the haploid mouse genome (the genetic program in the sperm or egg) has 20 chromosomes; the human has 23. This reflects chromosome breakage and fusion events that occur periodically in evolution. The differences between our two species that have been accumulating for 70 million years make it easy to overlook the more impressive similarities that have been retained. Yes, we live much longer than mice do, and we are a lot bigger and (probably) a lot smarter. But we share the same general body plan and the same organ systems. At the cellular level, it is difficult to tell us apart. If you look at a smear of mouse blood under a microscope, you will not be able to distinguish it from human blood.

Where it really matters—at the genome level—mice and humans are quite alike. Mice and humans each have about 30,000 genes, and most of them are fundamentally similar. The mouse genome is about 12% smaller than the human (2.5 versus 2.9 billion DNA letters), but very little of the extra human DNA codes for functional genes. Mice do have many more genes devoted to sensing odors than do humans. They also seem to have a lot more genes that directly influence mating behavior. About 90% of human genes that have been associated with disease have counterparts in mice, and they probably pose similar risks. The X chromosomes of humans and mice carry roughly the same genes, albeit sometimes in a different order. Comparing whole genomes, there are about 180 long blocks of DNA (up to 40,000,000 base pairs) in the mouse that contain genes which are lined up the same way and virtually identical in function to similar stretches of DNA in humans! Such facts led some scientific wag to suggest that humans are just mice without tails. He was in part wrong. Humans do have genes to make tails; they just never turn on.

How can humans and mice be so similar, yet so different? The reason the two mammals are physiologically so similar is that nature is conservative; if evolution solves a physiological problem, that solution is used repeatedly through the eons until environmental forces acting on one or more mutations in relevant genes select for one or more new variants. The reason the species are so different is that during the 70 million years since they separated from a common ancestor, they have accumulated vast numbers of significantly different mutations, many of which (apparently because they make the species more robust) have driven the biological divergence. Assume that the species currently differ in 2% of their DNA. Because genomes are so large, that difference is more important than it may at first seem. Humans and mice have about 3 billion DNA base pairs in a haploid (sperm or egg) genome. Two percent of 3,000,000,000 is 60,000,000. That constitutes more than enough potential variation in coding sequence to anticipate a significant number of differences in protein function and phenotype.

Mice and humans have been socially intimate for millennia. Mouse skeletal remains are regularly found in the excavations of archaeological sites that are more than 25,000 years old. Just as we do, mice like warmth and food. Doubtless, mice were extremely pleased by the development of agriculture in the Fertile Crescent about 10,000 years ago. They probably began to pilfer grain almost as soon as humans learned how to harvest and store cereals. The scientific name *Mus musculus* derives from two ancient Sanskrit words: mus means "thief," and musculus means muscular. At least in the 18th century when mice received their Latin sobriquet, scientists thought of them as big thieves.

The mouse best known to North American readers—the house mouse—is actually an "Old World" mouse which probably arrived with the earliest European explorers and quickly found a niche that was not occupied by any of his many "New World" cousins (mainly, those in the genus *Peromyscus*). Over the last few centuries, mice and humans have had an uneven relationship. Although farmers regarded them as vermin good only for fresh cat food, pet fanciers around the world took to them long ago. In the 17th century, Japanese collectors prized albino mice and experts wrote texts on how to breed them.

The 19th century witnessed the birth of biomedical research. Small in size, highly fecund, easy to care for, and without a champion in the animal rights world, the mouse was a favored animal in which to perform bacteriological and physiological studies even before the birth of genetics. Pasteur, Koch, Virchow, and many others used mice in their labs. With the rise of Mendelian genetics and a growing desire among researchers to use the tools of genetic analysis to investigate the medical problems of humans, there was an obvious need for a model organism closer to us on the evolutionary tree than are fruit flies and corn. This need eventually gave birth to the field of mammalian genetics.

A favorite pastime of scientists is to debate who really deserves the credit for an important scientific development. There are several people who may deserve credit for the origin of mouse genetics. Among them, my sentiments lie with a schoolteacher named Abbie Lathrop. Forced into early retirement in the late 1890s by illness, Lathrop first tried her hand at raising poultry on her farm in Granby, Massachusetts. When that venture failed, she decided to try to breed and sell fancy mice for the then popular pet mouse market. Somehow, geneticists at Harvard heard about her work, and she was soon selling to them. At some point, Lathrop noticed lumps in the skin of some of her mice. Puzzled and concerned, she contacted a number of pathologists, but only one, Leo Loeb, a professor at the University of Pennsylvania, agreed to examine them. Loeb discovered that they had cancer, a finding that led to a long collaboration on the study of the heritability of tumors in mice. Lathrop eventually co-authored ten papers with Loeb.

In the early 20th century, studies of the inheritance of coat color in mice were among the first experiments undertaken in mammals based on Mendelian principles. The French zoologist, Cuenot, showed that the offspring of matings between an agouti (the name for the regular gray coat) and an albino always resulted in agouti young. However, when the brothers and sisters were mated, their offsping were of three colors: agouti: black:albino, which appeared with the predictable ratio of 9:3:4. (The explanation for this odd ratio is that two separate genes determine coat color and that albino mice have genes for black coats, but lack another gene that is required for coloration to be expressed).

Karen Rader asserts in her scholarly book, *Making Mice*, "American mouse genetics was born at Harvard's Bussey Institution," then one of the few funded research centers in the United States. As early as 1900, William Castle, a Harvard professor of zoology (and one of Lathrop's first customers), was studying the genetics of several diseases in mice. Castle's mouse breeding (the word genetics did not come into existence until 1906) program was not large, but it was probably the first academic effort to create and maintain highly inbred strains. In addition, it drew graduate students who would transform the field. No one had a greater impact than did Clarence Cook Little, a direct descendant of Paul Revere, an outstanding student-athlete at Harvard, and the father of "big" mouse science in the United States.

Little had an unusual career. After earning his Ph.D. at Harvard, where he developed his lifelong interest in using highly inbred lines of mice to dissect the genetic contribution to disease, he spent four years doing research at the Station for Experimental Evolution at Cold Spring Harbor. Then he became—at the tender age of 34—president of the University of Maine (at the time a small, immature, and insecure institution). From the start, there was tension with the trustees (who were trying to build the overall educational quality of the school rather than focus on making it a haven for science) because of Little's desire to do research while discharging his duties as president.

During the second summer of his presidency, Little met a number of influential Detroit businessmen who vacationed in Bar Harbor. They must have been extremely impressed with him, for in 1925, just three years into his tenure at the University of Maine, Little was recruited to become president of the much larger and more prestigious University of Michigan. Little moved to Ann Arbor, convinced that he could build a major mouse genetic research program and guide the tiller of a well-established, large university. Research funds came with his appointment, and for a while it appeared that the driven young scientist could competently execute two full-time jobs. Over the next few years, however, his relationship with the faculty deteriorated, and Little finally realized that he was much more interested in being a scientist than a university president. During the mid- to late 1920s, he convinced wealthy Detroit patrons that the future of medicine depended on good science, that human health care (especially the struggle against cancer) would greatly benefit from a major effort in mammalian genetics, that the mouse was the right experimental organism, and that he was the right man to lead a bold new initiative in mouse genetics at a lab to be set up in Bar Harbor, Maine (the favorite summer resort of some of the patrons).

The Roscoe B. Jackson Memorial Laboratory (named for one of the first patrons, who died unexpectedly) was incorporated on May 4, 1929, an event which, in those less technological times, the *New York Times* reported three days later. Given that the success of the new institute was closely tied to philanthropic support, the new director's timing could not have been worse. The lab opened for business in early October, and the stock market immediately crashed. Promised gifts began to shrink or disappear altogether. The first two decades of the laboratory's existence were dominated by a constant search for operating funds. Nevertheless, the small staff did remarkable work in developing and characterizing new strains of mice and using inbred lines to study cancer. The hardest choice (one that was necessary for financial survival) was the decision to operate a side business as a supplier of inbred mice to other laboratories.

On October 23, 1947, the Jackson lab burned to the ground; all but a few hundred of the tens of thousands of specially bred mice that had become a staple of mammalian cancer research perished. Fortunately, the scientific community, government agencies, and philanthropists hastened to assist Dr. Little in rebuilding the institution. Scientists from around the country who had bought the JAX mice returned breeding pairs so that the colonies could be recreated. Given the fecundity of mice (about six litters a year), it was not long before there were again tens of thousands of extremely well characterized mice (many with special mutations that conferred risk of cancer and other diseases) flourishing on the coast of Maine. Thus, the lab was well poised to play a major role in the explosive growth of biomedical research that began in the 1950s and continues unabated.

Today the mouse is a key model organism in scientific research. The Jackson Memorial Laboratory provides researchers with about 2,000,000 mice a year, and at least that many are provided by two companies that provide similar services. The most widely studied of these are known as C57BL mice. This strain arose from a mating between Abbie Lathrop's mouse number 57 with her mouse number 52. She provided the offspring to Clarence Cook Little in 1921. We may know more about the biology of the C57BL mouse than of any other mammal in the world! By one count, more than 100,000 research papers have been published based on work with these little guys.

The next time you consider buying traps to kill the vermin who live in your basement, think for a moment about the astounding list of contributions that their relatives have made to human health. Taking a page from David Letterman, you might consider the following as "The Top Ten Reasons You Should Not Kill a Mouse."

During the 1920s and 1930s, Maude Slye, a researcher at the University of Chicago, undertook the Herculean feat of creating inbred strains of mice to prove the existence of genes that predispose to cancer. Working alone for most of her career, Slye, who became known as "the cancer lady," conducted painstaking autopsies—the only means to determine definitively whether a cancer had developed—on about 500,000 mice! Slye's mouse research taught us that mammals can be born with a genetic predisposition to develop cancer. It opened up avenues of research that led to the development of DNA-based tests to screen for risk for breast, ovarian, and colon cancer in some families today. Almost certainly, many more such tests will be developed in the next decade.

In 1949, a year in which 50,000 Americans developed acute poliomyelitis, mice played a key role in the fight against that dreaded disease. Three Harvard scientists, John Enders, Thomas Weller, and Frederick Robbins, showed that they could propagate the polio virus in a medium other than nerve cells. They proved it by injecting the fluid from the growth medium in which they had cultured a strain of the virus known to infect mice into the brains of healthy animals. The inoculated animals rapidly died of polio. This work was a crucial step to the development of the polio vaccine. Since 1954, polio vaccine has averted many millions of cases of the disease. Thanks in part to mice, today the world is on the brink of completely eliminating this horrible illness.

During the 1950s and 1960s, embryologists and geneticists used mice to explore the mysteries of germ-cell development. John Biggers, Ralph Brinster, and Anne McLaren learned how to coax immature mouse eggs to develop outside the body and to transfer them into the uterine horns of pseudo-pregnant mice. This paved the way for the successful use of in vitro fertilization to overcome infertility in humans. Building on the mouse work, Robert Edwards and Patrick Steptoe were able to develop techniques that led to the birth of Louise Brown, the first "test tube" baby, in the United Kingdom in 1978. In 2004, more than 20,000 test tube babies were born in the United States alone. Mice have helped hundreds of thousands of couples to have children.

In 1962, researchers discovered the "nude" mouse. In addition to being hairless, this mutant is born without a thymus gland and lacks a proper immunological defense system. Thus, it is possible to transplant a tumor from another animal (including humans) into the peritoneal cavity of the nude mouse to study both tumor growth and the antitumor activity of experimental drugs. Over the last 40 years, thousands of such studies have been undertaken. The nude mouse is a major tool in the study of cancer and transplant immunology. Many cancer survivors are alive today because of drugs first studied in mice. In 1983, researchers discovered a mouse with severe combined immune deficiency (SCID). Like nude mice, SCID mice are a valuable medium in which to study the growth and behavior of tumors. They also provide a superb natural cell culture chamber in which to grow hybridomas, cells that make a continuous supply of antibodies. These antibodies are used for a wide variety of diagnostic and therapeutic purposes.

In the late 1980s, it became possible to create "transgenic" mice. Scientists learned how to microinject a foreign gene into early mouse embryos in a manner that resulted in some of the genes being incorporated into some of the embryos. This opened up an entirely new approach to studying disease. In 1985, Harvard scientist Philip Leder and his colleagues became the first group to win a patent on a transgenic animal. In its genome, the so-called "oncomouse" carried a human gene variant that strongly predisposes it to develop breast cancer. The Harvard mice quickly became an important resource in the effort to develop new drugs to treat cancer. The award of the oncomouse patent triggered an ongoing international debate about the ethics of "patenting life." In the United States, the law is clear. It is possible to win utility patents on life forms that have been transgenically altered to accomplish a specific purpose.

Since the mid-1990s, the mouse has become an important organism in which to study the genetic control of the life span. Simply selecting for inbred mice that are long-lived, and back-crossing them, scientists have been able to create strains that consistently live 50–100% longer than the average mouse. To study the relationship between body size and life expectancy, scientists have bred unusually small mice. They have also created mice deficient in an enzyme called telomerase that is thought to play a key role in ordering cell division. Similarly, they have created strains with slow cell division times, a status that correlates well with long life. The knowledge to help humans live robustly for a century or more will come in part from studies of mice.

In 1999, Princeton neurobiologist Joseph Tsien briefly captured world attention when he and his colleagues announced that they had created a transgenic mouse that exhibited superior ability in learning and memory tasks. Dubbed "Doogie, the Smart Mouse" (after a character on a then-popular TV sitcom based on the adventures of a boy genius who becomes a doctor while still a teenager), the mouse carried an extra copy of a gene that codes for one of several receptors of a neurotransmitter called NMDA, a protein long believed to play a crucial role in the biochemical events that create memories. It appears that increased levels of the protein (called NR2B) in effect give the brain more time to create the memory tracts. In writing about Tsien's research, the popular press could not resist speculating that this "brainier" mouse represented proof of principle that we will eventually be able to modulate brain chemistry to make people smarter. Does the current scandalous use of performance-enhancing drugs among professional and amateur athletes reflect a culture in which drugs designed to enhance intellectual performance will find a willing clientele? No such compounds yet exist, but someday they may. I have little doubt that among students searching for an edge in the SATs there would be willing users. On the brighter side, it may be that mice like Doogie will help us find new approaches to ameliorating chronic neurological disorders such as Alzheimer's disease.

In December of 2002, researchers announced that they had produced the full (2.5 billion base pairs) sequence of the mouse genome (the second mammal—after humans—to be finished). The availability of the full sequence of the genetic "blueprints" for humans and mice has opened up broad new lines of scientific study, especially in the area of comparative genomics. For example, one group has already found evidence to infer that about 2% of the genes that were in the genome of the last ancestor common to both mouse and man have been deleted in the mouse. Similar studies show that over the last 70 million years there have been hundreds of small changes in gene order (chromosomal evolution) between the two species. Another important consequence of the completion of the Mouse Genome Project has been a rapid proliferation of projects to create new, highly defined mutant strains of mice that will greatly increase our understanding of gene function. At least ten such projects are under way. The information they generate could significantly enhance efforts to develop new drugs.

Ironically, the Mouse Genome Project (which focused on a highly inbred strain) has rekindled interest in wild mice. The many strains of naturally occurring wild mice around the world collectively carry thousands of genetic variants. When they are mated to the highly inbred strains for which we have the full sequence, the offspring become excellent models in which to study how slight genetic changes influence risk for disease.

Thanks to the work at Jackson Memorial Laboratory, scientists have access to an ever-growing number of inbred strains of mice with genetic disorders that mimic human disease. The JAX collection includes mice that are deaf, mice that are blind, mice with birth defects ranging from cleft palate to congenital heart disease, mice with bone disorders, mice with hereditary cancers, and mice with autoimmune disorders. Thanks to techniques that allow controlled creation of mutations, the "new mutant" collection is growing steadily. Today, the intellectual descendants of Clarence Cook Little annually provide more than 2,000,000 mice from among 2,800 inbred strains to more than 12,000 research labs located in 63 countries.

Now in its second century as a research servant to humanity, the scientific star of the mouse shines brightly. In 2006, some highly inbred mouse astronauts will orbit Earth inside a spacecraft programmed to spin in a way that generates a gravitational effect equal to that on the surface of Mars (0.38 of earth gravity). Each mouse will have her own habitat, including an exercise tube, the entirety of which will be under surveillance by a tiny Webcam! All of the mouse astronauts will be female; in an environment where every ounce of payload costs lots of money, there is a double advantage to females: They eat less than male mice, and they appear more sensitive to the effects of weightlessness. After five weeks in orbit, the mice will (collectively) parachute to earth, landing near Woomera, Australia. Then, they will be flown (first class!) back to the lab for physiological studies, especially in regard to loss of bone mass. These experiments in weightlessness may provide us with new insights into the risk for osteoporosis.

Virtually every American child who grew up in the 1950s was a fan of Walt Disney's TV show, "The Mickey Mouse Club." What the fans did not know is that a quite different kind of mouse club had existed long before Disney's. In the 1920s, as part of his effort to develop mouse genetics, Clarence Cook Little founded the "Mouse Club of America." The members—all mouse geneticists—corresponded regularly and met occasionally to fill each other in on advances in their labs. One of Little's best marketing ideas was his effort to convince Disney to support the Jackson Memorial Laboratory because his poor mice were relatives of the rich and famous Mickey. Little opened the Jackson lab the same year—1929—that Disney created Mickey. In 1954, the year Mickey made the cover of *Life* magazine, Dr. Little wrote to Disney asking for support. The great cartoonist and dream maker did not respond. Fifty years later, the Disney Mickey Mouse Club has receded from the cultural scene, but the scientific Mouse Club of America, a legion of mouse geneticists, continues to make extraordinary advances in its effort to use the humble mouse to improve human health.

Barbara McClintock in her cornfield at Cold Spring Harbor, circa 1940. (Courtesy of Cold Spring Harbor Laboratory Archives.)

Corn

Unlike most fruits and vegetables that humans eat, corn (also called maize, and I will use both words) provides nutrition from its seeds. The ears of sweet corn that one buys at a roadside stand in late summer have about 800 kernels. Each kernel is an immature seed, a germ cell engulfed by endosperm, a mix of carbohydrates and sugars that can sustain a new plant until it can support itself by photosynthesis. After appropriately modifying it by boiling and the liberal application of butter and salt, humans savor the endosperm.

For those who love fresh corn in late August, there is no end of stories praising local varieties. One of my favorites is the passion that people who live near the Eastern Shore of Maryland have for a variety called Silver Queen, a blissfully delicious white kernel corn that food writers have called the "sweetest imaginable." In reality, very little Silver Queen is grown anymore, and the reason is genetic. Silver Queen is wonderfully sweet, but its sweetness—driven by the "su" gene—rapidly fades as soon as it is picked. With the temperature well over 80 on an August afternoon in Maryland, at least half the sugar in the Silver Queen kernel is converted to starch in less than 24 hours. Simply put, Silver Queen has no shelf life. This led commercial seed producers to replace Silver Queen with hybrids (varieties called "se" or "sh2" for sugar enhanced or supersweet) in which the inevitable conversion of sugar to starch takes substantially longer. Today, roadside stands on the Maryland shore are loaded with white kernel corn that is very sweet, but it's not Silver Queen. There are many new varieties with similar names, but the purists will probably only get what they want if they find an heirloom plot—less than an acre just behind an old farmer's kitchen—and beg for that which is not for sale.

Corn is the world's third (after wheat and rice) most planted field crop. In 2007, farmers in the United States planted over 90 million acres of corn. Worldwide, the total acreage under cultivation approached 400 million acres. Most of the corn crop (75%) is grown in the United States, China, Brazil, and Mexico. Only about 10% (mostly varieties called flour or blando) of cultivated corn is eaten by humans. Nearly 75% of the planted acreage will grow strains called dent or dentado (their kernels have a slight indentation due to moisture loss) that will be used for animal feed and industrial purposes (such as the production of syrups and alcohols).

Maize (*Zea mays* ssp. *mays*) is a large domesticated grass. Its genome has 2.29 billion base pairs—about 80% the size of the human genome. About 70% of the corn genome is composed of noncoding repetitive DNA sequences, and less than 5% codes for proteins. Corn has more than twice as many genes (59,000) as do humans (~25,000). How can this be? With our big, complex brains, don't we need a lot more genes than corn? Apparently not! In its own way, corn is an extraordinarily complex organism. Photosynthesis is a complex process, and the corn life cycle requires many biochemical pathways that humans do not have.

The most distinctive characteristic of maize is that (unlike other grasses, each flower of which is bisexual) the sexual organs are separated. The male structure is the tassel at the top of each plant; the female structures are the ears (which are really condensed lateral branches from the main stalk). In the natural setting, fertilization occurs when breezes carry the male pollen from the tassel to the small female florets. When researchers want to study and control corn breeding in the field, they simply tie plastic bags over the tassels and use pollen of their choice to control fertilization.

The size and yield of a corn plant vary greatly, mostly as a function of genetics, climate, and methods of cultivation. In a temperate climate, corn plants spaced 30 inches apart usually grow to 8 feet and produce two to four ears. If the corn seed is sowed 15 inches apart, the resulting plants will be smaller and each will have on average only one cob. These limits have huge implications for third-world farmers trying to eke out an existence on a few acres or less. They also pose an important challenge to genetic engineers who are working to improve yield.

In both the Old World and the Americas, the rise of civilization was tightly connected with the development of agriculture. In the Old World, humans were fortunate to have relatively ready access to a number of animals (cattle, pigs, goats, sheep) that they domesticated and which relatively soon came to provide them with—according to some estimates—as much as 30% of their calories. In sharp contrast, the archaeological record of the Americas indicates that tribes which inhabited the southwest derived less than 10% of their calories from animals. What animal calories they did consume came almost exclusively from hunting. Given the dearth of easily domesticated animals, the early American peoples were fortunate to have lived amidst a rich diversity of naturally occurring plants that they successfully cultivated. A partial list includes vanilla, chili, amaranth, chive, sunflower, quinoa, chocolate, at least four kinds of beans (lima, summer, tepary, and jack), pumpkins, tomatoes, avocado, guava, and papaya.

The most important crop by far was corn—so important that one could reasonably argue that it literally provided the energy that fueled early American civilizations. Working together in the 1940s and 1950s, archaeologists and botanists demonstrated that when Europeans arrived in the 15th century, Native Americans were cultivating more than 150 varieties of corn in climates ranging from the cold, short summers of the St. Lawrence River valley in Canada to the hot, wet southeastern regions that we know today as Florida, to the high, arid plateaus in present-day Chile.

Humans and corn have been evolving together for as much as 6,000 years (the dating has large error bars). Although it is exceedingly difficult to trace precisely the evolutionary history of today's corn, the best guess is that it passed through two genetic bottlenecks. During the first—a series of human-driven crosses of the grass, teosinte (*Zea mays* ssp. *Parviglumis*)—the evolving species lost some of its genetic diversity (an inevitable consequence of initial domestication). The second and much more recent bottleneck involved much greater focus on specific physical traits (a process known as artificial selection). Maize geneticists estimate that selection pressure on just 1,200 of corn's 59,000 genes led to the highly modified plant upon which people throughout the world now so heavily depend.

The corn we eat today probably evolved from innumerable fertilization events involving varieties of teosinte that were growing wild in the Sierra Madre of western Mexico at elevations of from 1,000 to 5,000 feet.

Archaeologists have found evidence that teosinte was first domesticated in a region known as the Balsas River drainage area about 4000 B.C. There is ample evidence that 1,000 years later (about 2700 B.C.), early farmers were growing a small-eared maize that had only 6–9 kernels per cob, and that they were grinding those few kernels to make flour. This was probably an ancestral "pop corn," so named because there is enough moisture within the endosperm that upon heating it will vaporize and burst the shell. Why were early farmers so dedicated to such a poorly producing plant? They must have grasped its nutritional value. When ground to a dry flour, maize kernels yield 3,600 calories per kilogram. Corn is one of the cheapest sources of high-calorie nutrition on the planet. A corn flour tortilla may rank second only to rice as the most common food on humanity's table.

Archaeological research suggests that maize has been the single most important source of calories in Central America for more than 3,000 years. By 1400 B.C., many local varieties of maize could be found at any settlement in any region of what we know today as Mexico. The Spanish colonists quickly recognized the nutritional value of corn. Within two centuries after their arrival, the Spanish had distributed corn around the globe, and it was being grown almost everywhere. By some estimates, the citizens of modern Mexico still eat on average a pound of corn flour every day. In rural areas, maize is the source of as much as 70% of daily calories.

Maize has been the key crop in the lives of Meso-American peoples for so many millennia that it plays a central role in various creation stories. In the accounts of the Mayans, the gods did not succeed in making men until they mixed their own blood with maize dough. In some Meso-American languages, the word for maize flour is translated as "our flesh." The word "teosinte" may be translated as "divine grain." According to a 16th-century Spanish priest, the Aztec people held maize in such reverence that if they saw kernels on the ground they would quickly pick them up, thinking it was an offense to their deity not to gather up what was so essential to their sustenance. The Meso-Americans were not herders, and they derived almost all their protein from plants. Eaten together, maize and beans have a protein content similar to milk.

Worldwide, corn may be only the third largest crop, but it is the most ubiquitous. Centuries of coevolution with man have resulted in strains that can grow in regions as cold and dark as those near the arctic circle, as dry and hot as the deserts of the American southwest, and as wet as some of the rainiest parts of southeast Asia. Around the globe, mature corn plants range in size from 2 to 16 feet. Depending on location and cultivation, strains may mature in as short a time as 60 days or as long as 330. Some strains of corn can be grown at an altitude of 12,000 feet. Depending on the strain, corn plants yield from one to four ears. Each ear may have as few as 10 or as many as 1,500 kernels. Agricultural yield can be as small as half a ton to as many as 24 tons per hectare (10,000 square meters or 2.47 acres). The kernels can be of many different hues, ranging from bright red to a lustrous white. The Mayan people, who have the most longstanding and deepest connection to maize, believe that the colors of the kernels reflect a creation story. In their history, maize became available to humans when a god threw a thunderbolt and broke open the mountain in which the seeds were stored. The color indicates how close the seed was to the fiery dart; black seeds were scorched, while white kernels were completely spared.

Today there are hundreds of different strains of corn under cultivation in the world, but in the vast majority of cases, each is found only in a single, small area. Much of the natural genetic diversity of corn is perpetuated by small groups of farmers in Mexico and elsewhere growing strains that have over hundreds or thousands of years been adapted to a local geographic niche. Given the rapid industrialization of Mexico, this diversity is under inexorable pressure that conservation efforts can at best slow. Experts who have studied corn growing in rural Mexico estimate that in 2000 there were only 42 races (varieties) under cultivation, down from perhaps 300 several centuries ago. If the decline continues and only a few races remain under cultivation outside of agribusiness farms, the world will lose an important biodiversity reserve.

In the United States, the value of the American crop approaches $30 billion and is an important source of export revenues. Over the last five decades, major corn producers have put great emphasis on creating seed that yield maximum economic benefits. For that reason, most of the world's corn is today produced using substantially less than 5% of its known genetic diversity. These highly uniform crops may carry with them a growing vulnerability to a major ecological disaster (genetic diversity tends to limit the devastation wrought by a plague). Future breeding studies of corn will have to rely more and more on the study of exotic, wild strains as a source of gene variants that create enhanced nutritional value and maintain pest resistance.

Scientific breeding to develop valuable new strains of corn has a venerable history. For example, without interruption since 1896, scientists at the University of Illinois have continuously selected corn seed on the basis of its oil content. For 109 years, scientists have successfully increased the average oil content from a starting point of about 5% to yields that currently exceed 20% in the richest strain. Efforts to produce a low-oil corn have been equally successful. In 2005, the current scientific team at Illinois reported painstaking work that focused on trying to identify the gene variants responsible for the gradual improvement in oil yield over 100 generations. They concluded that about 50 different genes, each of very small overall effect, were involved in the enhancement of oil content, and they identified about two-thirds of them. The data indicate the genes have "additive" effects, meaning that they act independently. Besides its agronomic value, the University of Illinois maize study can be appreciated as a human version of natural selection. It is the action of the environment (in this case, humans) on myriad gene variants of small effect that has driven change in the species.

A few years after Thomas Hunt Morgan and his students in the "fly lab" at Columbia University made history by constructing the first genetic map of *Drosophila melanogaster*, R.A. Emerson, a newly arrived professor of plant breeding, began developing maize genetics at Cornell. Emerson's work on the inheritance of plant colors placed the genetics on a firm foundation. The revolutionary advances in understanding started in the late 1920s when a young, iconoclastic Cornell graduate student named Barbara McClintock set out to do with maize what Morgan and his team had done with fruit flies—define each of the plant's "linkage groups" and correlate them with its ten pairs of chromosomes. This project, which she successfully completed over the next few years, created a bridge between genetics and breeding and gave birth to the field of plant cytogenetics (the study of chromosomes).

Looking back from the vantage point of several decades, Marcus Rhodes, in his day one of the world's most famous plant breeders, called the period from 1929 to 1935 at Cornell (where he also worked) the "golden age" of corn genetics. Many of the discoveries made there, such as of the genetic control of chromosomal behavior and the consequences of non-homologous pairing of chromosomes (which permits gene exchange), are fundamental to all organisms. McClintock's most far-reaching discovery—of a phenomenon called transposition ("jumping genes")—was not appreciated for decades, but it was cited when she was belatedly awarded a Nobel Prize.

GENETICALLY MODIFIED CORN

For more than a century, advances in genetics have periodically spurred acrimonious public debate. The eugenics movement of the early 20th century, the race/IQ debate of the mid-century and beyond, the public fright over the potential dangers of genetic engineering of bacteria in the 1970s, and the sociobiology controversy of the 1970s and 1980s are among the best known. For the last five years, there has been a bitter debate over the permissible scientific uses of human embryonic stem cells with the contending factions polarized along force lines drawn decades earlier by the abortion controversy.

For more than a decade, corn has been at the center of a passionate and often vitriolic public debate concerning the transformation of commercial agriculture by the rapid and widespread adoption of genetically engineered crops. Opponents of genetically engineered crops are legion, especially in Europe. Three of their major concerns are: (1) genetically altered crops could be unsafe, (2) agribusiness is behaving with Promethean hubris in its cavalier manipulation of the natural world, and (3) the business practices of those companies that control the world's genetically engineered seed will harm the economic prospects of tens of millions of third-world farmers. I side with those who favor the rapid, safe development of genetically modified groups. I think it could be among the most important positive events for the environment in history.

The development of a stable, highly efficient agriculture was a prerequisite for the emergence of civilization. Agriculture is in essence the selection and domestication of both plants and animals that are then enhanced for characteristics beneficial to mankind by use of breeding practices that drive evolutionary change far more quickly than does natural selection. By any measure, plant and animal breeding has dramatically altered many species. If they could see a typical ear of corn today, the Meso-Americans who first domesticated it might not even recognize that it is corn. Why then do consumers in England, farmers in France, and dockworkers in Germany perceive genetically engineered corn to be different from all the other modifications of corn that man has undertaken? Do they really grasp the scientific nuances and distinguish between the two practices? Is there really a fundamental safety difference in the fact that, unlike crops developed by classic breeding, genetically engineered (transgenic) varieties are created by moving one or more genes across a species barrier?

The product that became a major target of those opposed to genetically engineered crops is called Bt corn. The letters, Bt, refer to a protein produced by the bacterial gene found in *Bacillus thuringiensis*. This protein is extremely toxic to certain insect pests, such as the corn borer beetle, which can devastate crops. In the late 1980s, scientists figured out how to use a plasmid (a small circle of DNA) called ti from another bacterium called *Agrobacterium tumefaciens* to act as a vehicle that could transport other genes into the corn genome. After fusing the Bt gene to ti, the scientists incorporated the DNA sequence (they use the term, transfected) into the corn germ plasm. As hoped, the plants that grew from Bt seed were remarkably resistant to insect pests. Since economic losses due to the corn borer beetle were known to approach a billion dollars a year in the United States and Europe, there was a huge economic incentive to grow crops from genetically engineered seed.

The first transgenically created seeds of major crops became available in 1995. The positive economic impact (as measured by higher yield per planted acre) of products like Monsanto's YieldGard corn was immediate and dramatic. In 1995, 4 million acres of genetically engineered corn were placed in cultivation; by 1999, there were 200 million acres planted with transgenic seeds. In just a few years, industrial plant scientists had created similar resistant strains of soybeans, cotton, rice, and other plants. At the same time, they were embarking on research to develop strains in many other commercial crops, ranging from yellow squash to papaya, to overcome the most burdensome pests they encountered.

The rapid commercialization of foods made completely or in part from genetically engineered corn evoked a sharp, sustained, and largely unexpected attack, mostly in Europe. In both the United States and Europe, a 100-year-old American company, Monsanto, was at the center of the controversy. (The Swiss firm, Ciba-Geigy, now Novartis, was actually the first to launch a genetically engineered corn product in 1996, but perhaps because it was far better known as a pharmaceutical business, it did not become the lightning rod.) As far back as the mid-1980s, Monsanto, already committed to a future in biotechnology and sensitive to potential public concerns, had adopted what its executives viewed as a cautious approach. For example, in a 1986 visit to Vice President Bush, Monsanto officials urged that Washington write regulations on the production and marketing of genetically engineered crops. Monsanto thought that when it launched its product, the claim that they complied with government regulations would reassure the public. The administration complied and in 1992, during the first Bush presidency, Vice President Quayle spoke in favor of the nascent industry.

In 1995, a new CEO, Robert Shapiro, took the helm at Monsanto. Almost overnight, the giant corporation changed its strategic direction. Convinced that biotechnology could simultaneously improve the efficiency of world agriculture and make Monsanto fabulously successful, in the late 1990s, Shapiro bought several potential competitors and cut aggressive research deals with lead companies in the emerging genomics industry. Shapiro also abandoned the cautious approach to product launch of genetically engineered seeds in favor of an aggressive approach. He was convinced that the evidence on product safety was so strong that public fears would not be substantial. In so thinking, he was largely correct about consumers in the United States, but with regard to Europe he was wildly off the mark.

The change at Monsanto played into the hands of dogged critics like Jeremy Rifkin at the Foundation for Economic Trends (who had been sounding alarms about genetically modified crops since 1992). Over a couple of years, Rifkin galvanized a wide coalition of national environmental groups to question the speed with which agriculture was being transformed. In Germany, the politically influential Green Party virtually halted the importation of genetically engineered crops. Its major concern was that no one could reassure the public that an unknown ecological hazard of grave proportions would not arise from the spread of transgenic crops. In France and Italy, small farmers rallied against them. Their major concern was that genetically engineered corn and other crops would harm their effort to survive in a market dominated by a few large agricultural companies. During the period 1995–1999, United States exports of corn to Europe, a $400 million market, almost disappeared.

A key argument made by opponents of genetically engineered foods was that the pre-market assessment of environmental risk had been woefully inadequate. Thus, in 1999, when a young Cornell scientist named John Losey and two colleagues published a brief paper in *Nature* reporting experiments suggesting that pollen from Bt corn might kill countless millions of monarch butterflies, they landed on the front page of the world's major newspapers. Critics saw the threat to the lovely monarch as proof of the latest example of capitalist hubris. Some read the scientific paper as a warning that humans should not be moving genes across species barriers. What would happen, they mused, if pesticide-resistant genes entered the genomes of important weeds?

The issue was so hot that a dozen or so laboratories began to conduct similar experiments. On September 15, 2001, six different groups of scientists, each of which had examined questions related to Losey's work, published papers in the *Proceedings of the National Academy of Sciences* concluding that Bt corn posed no risk to monarch butterflies in the wild. The papers calmed, but did not end, the debate. So the Department of Agriculture, the FDA, and the EPA commissioned the National Academy of Sciences to study the safety issue in depth. In July 2004, it issued its report, which concluded that genetically engineered crops posed no risks that required a special level of oversight. Ironically, the scientific studies that partially vindicated genetically engineered crops came too late to save Monsanto. Having suffered severe economic harm from the debate over "Frankenfoods," as they were called in Europe, in December 1999, a few months after Losey's paper appeared in *Nature*, Monsanto agreed to be acquired by Pharmacia & Upjohn (a pharmaceutical company that has since been acquired by Pfizer).

The safety debate, which was most intense in Europe and generated government-imposed moratoriums on the introduction of genetically engineered crops from Brazil to Indonesia, crested in 2000–2001. In 2004,

Brazil gave the green light to a burgeoning array of genetically engineered crops, and even Britain approved the planting of one strain of transgenic corn. But in Africa, the continent that could most benefit from genetically modified (GM) food, many governments have been highly suspicious of taking food aid in the form of genetically modified corn. Of six nations in southern Africa gripped by severe drought in the winter of 2002–2003, two (Swaziland and Lesotho) accepted U.S. corn, but three others (Mozambique, Malawi, and Zimbabwe) only accepted U.S. corn that had been milled to flour, forbidding any use of the seed for planting. The sixth, Zambia, rejected genetically modified corn even though more than 3 million of its people were thought by international relief agencies to be at risk from death by famine. The Zambian government broadcast ads saying the corn was poison. Government officials, in part influenced by groups like Greenpeace, feared that the American GM corn contained new allergens and that widespread use would make farmers dependent on American biotech companies for seed.

The politics of GM food has spilled into many other African states. Uganda, a nation that is highly dependent on bananas (the average person eats 500 a year) is also in the midst of a food crisis because an airborne fungus called Black Sigatoka has destroyed or threatened well over one-half of that crop. In Belgium, a scientist has developed what may well be a solution, but it involves genetic engineering of banana seeds. Ugandan officials are in a quandary. They are torn between statements from the United States that GM food is safe and those from Europe that urge extreme caution. The problem gets ever more complex. American companies have genetically engineered a cotton seed that could increase crop yields in Africa, but European states that constitute the main market to which Africa exports have threatened not to buy GM cotton from them if they grow it.

As far back as 2001, the United Nations issued a report entitled "Making New Technologies Work for Human Development," which firmly advocated the planting of GM rice, millet, sorghum, and cassava throughout the developing world to combat malnutrition in more than 800 million people. On average, the strains in question have 50% higher yields and mature 30–50 days faster than current staples. They are also much richer in protein, far more resistant to pests, and better able to tolerate drought. Unfortunately, there are no genes to counter political posturing or overcome fear generated in Africa by well-meaning, but scientifically ill-informed, activist groups.

I think that harsh regulatory barriers to genetically modified corn and other crops will soon end. As I show in the next chapter, extremely wide use of genetically engineered rice in China is likely to be the key to turning the tide. But the change is already apparent. European Agriculture Commissioner, Mariann Fischer Boel, recently suggested that although GM crops should be segregated from non-GM crops in the field, regulations should not be so burdensome as to keep GM crops from the marketplace. This is a significantly more balanced position than the absolute bans that have been proposed in past years. India has approved the cultivation of genetically modified cotton in its fertile northern regions, permitting farmers to plant six varieties of seed that use Monsanto technology.

In 2003, about 40% of the U.S. corn crop comprised genetically engineered corn hybrids. In 2008, GM corn accounted for more than one-half of all corn acreage. Similar trends are under way for many other crops. Cornstarch, corn oil, corn syrup, baking powder, pastry fillers, even some nutritional supplements—all may contain traces of GM corn. For you moviegoers, there is not yet any GM popcorn on the market. In all the billions of times that humans have consumed a product made with Bt corn, there is no evidence that it has harmed a soul. When one considers the terrifying threats inherent in our most serious environmental problems, such as hydrocarbon emissions and global warming, air pollution, and water pollution, those who have protested Bt corn appear to have directed scarce political resources at the wrong problem.

In addition to increasing yield and reducing the use of pesticides, the tools of genetic engineering can be used in biofortification—the alteration of plant metabolism to confer some nutritional benefit, such as increased concentrations of vitamins and minerals. Millions of children suffer severe illness due to inadequate supplies of vitamin A. In the next chapter on rice, I will recount how a few scientists are enriching cereal to have much higher concentrations of its precursor (beta-carotene). In 2008, a group of scientists from Cornell and elsewhere announced they had discovered a way to greatly increase the amount of provitamin A in corn by selecting for a variant of a key gene that codes for a protein called lycopene epsilon cyclase. The variant version drives the production of sharply higher concentrations of beta-carotene. They did not use genetic engineering to achieve this feat, but the technology might be helpful in building upon it. If these advances can be made widely available, the resulting corn could play a major role in reducing childhood blindness.

A handful of golden rice. (Photo courtesy of Monsanto Co.)

Rice

On average, each of the 6.5 billion people on earth consumes nearly 200 pounds of rice a year. In North America and Europe, the per capita consumption is low. Americans each eat about 20 pounds of rice a year. But in China, the average is over 200 pounds. The leader is probably Mynamar (formerly Burma) where on average each man, woman, and child eats nearly 250 pounds of rice in just one year. World rice production has been rising steadily for as long as statistics have been kept. Since 1961, human consumption of rice has risen from 200 million tons to more than 500 million tons (but still has not kept pace with world population growth). Patterns of consumption are not uniform. Over the past few decades, Japan sharply reduced its consumption of rice, while the consumption of rice in Africa almost doubled and will continue to increase.

Overall, rice provides one out of every five calories that humans consume. Rice is the major single source of calories for about one-half the people on the planet. From a historical perspective, rice, the cultivation of which was widespread earlier than wheat or corn, has provided more calories to humans than any other food. In North America, it is only 3%, and in Europe, less than that, but in Asia, rice currently accounts for 31% of all calories, and in Africa, it is 8% and rising (up from 5% just 40 years ago). In much of Southeast Asia, humans get much more than one-half their calories from rice. Wheat and rice are the two most important foods on the planet. A dramatic reduction in the production of either would kill millions of people.

Rice is a semi-aquatic annual grass. It is a summer crop that prefers temperatures between 65 and 90 degrees Fahrenheit. Most rice is sown in heavily irrigated land with the seeds taking root under water. It takes more than 1,000 gallons of water to produce two pounds of rice. The seedlings emerge through the water in about two weeks, and the plant grows rapidly over a 60- to 90-day period. The young plant then enters its reproductive phase, developing a structure called the panicle (the flowering head), which is the source of the buds. This is the stage at which pollination takes place. In general, depending on the strain, at this stage in their life cycle the plants are maintained in about eight inches of water, a procedure that protects from low temperatures. As the plant further matures (the ripening stage), the rice kernels (seeds) that humans eat develop. An adult rice plant is about three feet tall.

In Asia, the traditional method to harvest rice is to use a sickle-shaped instrument to cut the plants. The farmers then tie them in bundles and let them dry. They then carry the bundles to the village, where they whip the plants against a screen through which the dry kernels fall. In the United States and other highly industrialized countries, rice farmers use combines to cut and thresh the plants. They then dry the moist kernels in heated sheds.

A grain of rice has three layers—the husk, the bran, and the germ—that surround the endosperm. People cannot digest the husk, which is removed by a process called milling. The bran and germ layers are brown. Brown rice takes substantially longer to cook and is chewier than white rice. By removing the bran and germ layers, one is left with the kernel—the white rice that most of us eat. The choice of white rice over brown rice is the triumph of esthetics over good sense. Removing the bran and germ layer sharply reduces the nutritional value of the food. Brown rice is rich in calcium, potassium, phosphorus, and several B vitamins. A kernel of white rice is essentially a tiny mass of carbohydrate. The major constituent, which may account for more than 25% of the kernel's weight before cooking, is a starch called amylose. Amylose is a long molecule with limited side branches that is composed mostly of linked glucose molecules. The kernel also contains much of a closely related, highly branched, molecule called amylopectin. Rice kernels contain virtually no fats, no salt, and no simple sugars. They contain some B vitamins, iron, potassium, and phosphorus, but lack most other vitamins.

By far the most widely cultivated species of rice is *Oryza sativa*. The best molecular evidence suggests that the most ancient ancestor of the races of modern sativa that encircle the globe arose in southeast Asia about 15–20 million years ago. For millions of years, the range of wild rice was defined by animals that had eaten it. Rice reached Australia from Asia about 15 million years ago when there was still a land bridge connecting the two continents. Until about 3 million years ago, it was still relatively easy for animals to cross the Himalayan mountain range (at that time the mountains were less than half their current heights). After that time, the evolutionary history of rice in China and rice in South Asia began to diverge. Today, there are more than 50,000 varieties of rice on the planet. Most of them are wild or cultivated in a very restricted range. The vast majority of the rice consumed by humans comes from just a few strains.

The diaspora of cultivated rice is much more recent. Developed first in Southeast Asia more than 6,000 years ago, techniques for cultivating rice probably arrived in India before 3000 B.C. However, knowledge of rice cultivation probably did not reach southern Europe until the army of Alexander the Great carried it there about 320 B.C. The spread of rice cultivation in Europe was remarkably slow, perhaps in part because pools of standing water (in which it is grown) were associated with an increased risk for malaria! Rice reached North America in 1685 when South Carolina colonists began growing it.

The origin of rice, which may be the first crop that humans cultivated, is lost in antiquity. One measure of its long agricultural history is that in several Asian languages the words for rice and food and agriculture are synonymous. This is not the case for any other grain in any other language. The oldest archaeological evidence of rice cultivation was discovered at a site called Non Nok Tha in Thailand in 1966 when researchers found pottery shards with the imprint of both grains and husks of the rice strain, *Oryza sativa*, that could confidently be dated to 4000 B.C.

Both early Hindu and Buddhist scriptures make references to rice; early Jewish texts do not. Perhaps the best anthropological evidence of its extremely early appearance in human culture is that rice figures in the origin myths of many countries. Folklore in Burma holds that the first people (the Kachins) were sent forth from the center of the earth by the gods with seeds of rice and the promise that they would live in a fertile land where the grain would grow wondrously. In Bali, people believe that the Hindu god, Vishnu, created rice and that Indra taught people to cultivate it. In China, there is a saying that pearls and jade are not nearly so precious as seeds of the "five grains," of which rice is the foremost. In Japan, it was long believed that the emperor was the living embodiment of Ninigo-no-mikoto, the god who ripened rice. With rice providing one-fifth of the world's food, one might well consider it a gift from the gods. If so, the greatness of the gift is yet to be realized. For although rice sustains millions, it has the potential to provide much more.

In December of 2002, after nearly a decade of work, scientific members of the International Rice Genome Sequencing Project (IRGSP; a consortium of laboratories in ten countries) announced at a ceremony attended by the Prime Minister of Japan that they had completed a detailed first "draft" of the rice genome (which consists of about 390 million base pairs), making rice just the second plant (the first was a plant widely used in research called *Arabidopsis*) to have its DNA sequence so deeply analyzed. To accomplish this feat, all ten labs worked with the same variety of rice, using various biochemical techniques to chop its genome into about 150,000 pieces. The scientists amplified and sequenced each little piece. They then used computer programs to scan this vast amount of sequence data. The programs align the short segments by looking for sequences at each end that match sequences at the ends of other pieces. This exercise was repeated for each of the pieces ("clones") about ten times, ensuring that the work product was about 99% accurate. The IRGSP greatly benefited from the decision of two companies, Monsanto in 2000 and Syngenta in 2002, to share detailed physical maps of the rice genome that they had compiled. In August of 2005, the IRGSP announced that it had almost completely sequenced the entire 390 million bases (for technical reasons, some spots are much more difficult to decode than are others). The movement from a 99% accurate draft to a >99.9% complete sequence is a huge scientific advance.

The successful sequencing of the rice genome is a candidate for the title of the most important advance in human health in history. One can make this bold claim on two grounds. First, about one-third of humanity depends on rice to survive, so the inevitable increases in agricultural productivity that will flow from deeper understanding of the genome may do more to combat famine and malnutrition than any other event. Second, the rice genome is remarkably similar to the genomes of wheat, barley, oats, and other grasses, all of which are very important to human health. Knowledge of rice genes will almost certainly accelerate our ability to improve yields from other grains.

Of course, capturing the sequence is only the beginning. Genomes are complex and mysterious, and the deeper one looks, the more surprises one finds. Still, the current level of accomplishment is no small feat. Although the rice DNA sequence is much smaller than the human (we have 3 billion DNA bases; rice only has 390 million), rice has as many as 50,000 genes, roughly twice as many as humans. This is possible because humans have a lot more DNA that does not directly code for genes (think of it as spacer material) than does rice. With the DNA sequence of the 12 rice chromosomes in hand, scientists have launched an ambitious effort to find and study genes that are particularly important in maximizing the nutritional value of the world's most important genome.

Given that rice is the most important food in the world, it is no surprise that many scientists are trying to develop strains which are adaptable to climates far different from the wet, lush paddies of Southeast Asia. If, for example, one could develop strains of rice that would grow well in the cold steppes of central Asia and the harsh, arid soil of sub-Saharan Africa, one might greatly reduce the deaths from periodic famines. The two biggest obstacles to expanding rice production in much of Africa are the relative lack of water and the high temperatures. Starting in the 1980s, a Sierra Leonean rice breeder named Monty Jones began to tap into the knowledge of farmers about which local strains of African rice seemed best able to tolerate arid soil and a hot, dry climate. For more than a decade, Jones and his colleagues collected dozens of strains and crossed the most hardy but low-yielding African rice with the more fastidious but high-yielding Asian rice. By 2002, they had created several NERICAS (the acronym used for new rice strains for Africa). By 2004, farmers in West Africa had 30,000 hectares (75,000 acres) of these hardy, high-yielding strains in cultivation. In 2004, Monty Jones was a co-recipient of the World Food Prize, often called the equivalent of the Nobel Prize. The NERICAS promise both to increase local crop production and to reduce the need to import expensive rice from Asia.

Jones's work relied on classic plant hybridization techniques and required many years of effort. Professor Susan McCouch, a plant geneticist at Cornell University, who seems to log as much time on transoceanic flights as she does in the classrooms in Ithaca, New York, hopes to achieve even more impressive results by combining classic plant breeding with genetic engineering. McCouch partners with plant breeders the world over to search for strains of rice and other grasses which have variants in key genes that will allow them to adapt to harsh climates. Because the industrialization of world agriculture has greatly reduced the genetic diversity of our most important crops, her work, which focuses on breeding newly found wild strains of a crop with widely used commercial varieties, offers great potential to protect the rice plant's biodiversity. This reduces the chance that a hugely important species in the world food supply will be ravaged by a pathogen.

From her office at Cornell, McCouch runs one of the most ambitious academic networking systems imaginable. Her team works closely with breeders in Central America, West Africa, Asia, and the United States to investigate the potential for rapidly introducing useful genes from wild species that will improve grain quality, reduce dependence on moisture, or increase pest resistance. They have created a huge electronic database that attempts to capture, store, and share the immense amount of data being generated about the world's most important food. In her fieldwork, Professor McCouch crosses wild strains of rice with highly cultivated strains, then screens them for about a dozen traits (time to maturity, grain yield, plant height) under a variety of different soil and growing conditions. She then uses molecular genetic tools to map the genes that most influence the traits that are beneficial to the world food supply.

China is poised to effect a great change in world GM grain production. Chinese scientists have developed two strains of rice—each resistant to two insect pests, the stem borer and the leaf roller—that are now in the final phase of field-testing before commercial launch. Currently, hundreds of farmers in eight villages are growing the insect-resistant rice and comparing it to traditional strains. Early results suggest that although the GM rice only confers about an 8% increase in yield, the farmers' dependence on pesticides drops by more than 80%. There is a substantial possibility that, by 2007, farmers in China will be growing GM rice for human consumption on hundreds of thousands of hectares. If China opens the door to commercial production of these strains, it is highly likely that nations throughout Southeast Asia will follow suit. The widespread cultivation of GM rice could be the event that reverses political resistance to GM crops.

One of the many health risks to children in third-world countries is vitamin A deficiency, a disease almost unknown in the United States and Europe. Vitamin A (the letter reminds us that it was the first vitamin to be discovered, in 1915) is a generic name for several different chemicals that are derived from compounds called carotenoids. After a person ingests carotenoids (which are plentiful in milk, eggs, fruits, and many vegetables), the body changes them to compounds called retinols and retinals. Retinal is an essential part of the pigment called rhodopsin that is a critical component of the rods and cones, the cells in the eye that are photosensitive. Vitamin A also plays an important role in maintaining the stability of cell membranes in tissues throughout the body.

Children who have vitamin A deficiency disease have many different problems. They are usually retarded in their growth, and they often have anemia. Many affected children are well behind in their mental development, and they are often lethargic and listless. Researchers have shown that these children are at increased risk to succumb to infections like measles. Simply by providing at-risk children with twice-a-year injections of high doses of vitamin A, health care teams were able to reduce their mortality rate by 30% compared to that of a matched, untreated group of children. The most feared complication of vitamin A deficiency is blindness. Malnourished children are at high risk for two reasons. Because of damage to the rods and cones, the children first develop night blindness. Somewhat later, the damage to cell membranes causes the eyes to become unusually dry. Bereft of the protective moisture on the surface of the eye, the children suffer many small penetrating injuries and recurring infections and scarring of the cornea.

According to the Rockefeller Foundation, a leader in developing world agriculture, more than 100 million children suffer from vitamin A deficiency. The number defies belief, but it is not a misprint. The foundation estimates that vitamin A deficiency directly contributes to the deaths of as many as 2.5 million preschool children each year, about 30% of the world mortality in this age group. Vitamin A deficiency is the most common cause of blindness in children. Public health experts estimate that as many as 1 million children a year lose a significant amount of vision because they do not absorb enough carotenoids. The Rockefeller Foundation estimates that, each year, 500,000 children become completely blind. This terrible situation is especially sad because vitamin A is inexpensive and is available in pill form. The problem is that the children most at risk live— and die—in parts of the world where health care systems do not exist.

What has all this to do with rice? Nearly 20 years ago, Ingo Potrykus, a professor of plant physiology at the Swiss Federal Institute of Technology in Zurich, had the idea that it might be possible to use the tools of molecular biology to develop rice that is rich in carotenoids. Although rice is a grass, the endosperm (the grains we eat) has almost no vitamin A. Of course, plant breeders have long been aware of the tremendous health benefits that would derive from developing a strain rich in carotenoids. For many decades, they have hoped to stumble upon a wild plant with yellow kernels, the color that would signal the presence of high levels of carotenoids, but no naturally occurring mutant strain has ever been found.

Professor Potrykus joined forces with Peter Beyer, a geneticist at the

University of Freiburg in Germany, who had used daffodils to study the enzymatic steps by which plants make carotenoids. They found that rice plants lacked not one, but four, genes needed to complete the manufacturing steps. Using the tools of molecular biology, they were quickly able to transfer the daffodil gene (for phytoene synthase) that controlled the first of the four conversion steps into rice. One of the other key daffodil genes was very difficult to work with. Therefore, the scientists turned to bacterial genes that had the same biochemical duties. By 1999, they had created a strain of rice with two daffodil genes and one bacterial gene which produced enough vitamin A in its grains to raise hopes that further work would lead to a strain that would permit a child to obtain a significant fraction of the daily requirement of vitamin A by eating 300 grams (10 ounces) of rice a day.

Unfortunately, the "golden rice project" was quickly engulfed in the debate over the safety of foods into which genes from other species had been transferred. Ironically, the main opponents were groups based in Europe, a continent that does not experience chronic malnutrition or famine. Many of the opponents of genetically modified food were and are suspicious of the corporations that would control the world seed market. In addition, they (understandably) doubted that golden rice would solve a global malnutrition problem which was driven by many political and economic factors far more intractable than increasing the vitamin A content of the world's rice supply.

The controversy over the golden rice project reached its zenith just as scientists were completing the project to make a detailed map of the rice genome, a major step on the road to compiling the complete DNA sequence of its approximately 50,000 genes. In the spring of 2000, Monsanto, then one of the world's largest seed producers, announced it would give a "royalty free" license on intellectual property used to develop golden rice to an organization created by Potrykus to disseminate it. Monsanto also agreed to make public its first solid "draft" of the rice genome, a decision that greatly benefited the International Rice Genome Sequencing Project, an academic consortium working with substantially fewer resources. Unfortunately, the creators of golden rice used techniques protected by dozens of patents held by many different companies. Monsanto's generous act was only a needed first step.

The production of golden rice also depended on technologies that were protected by patents awarded to a large pharmaceutical company called Zeneca, which was not willing to be as generous as Monsanto. Potrykus and Beyer took the politically unpopular step of working with Zeneca rather than castigating it for enforcing its patent rights. They worked out an agreement with Zeneca that gave the scientists the right to provide golden rice to nonprofit rice-breeding programs intended to serve resource-poor farmers in the third world. The threshold, which defines as poor those farmers making less than $10,000 a year from rice farming, covers most of the population of Asia. Despite virulent criticism from a number of political groups, the staid Rockefeller Foundation has remained firmly committed to supporting the creation of a strain of golden rice that is rich in vitamin A.

In 2004, the scientific project to create strains of rice rich in carotenoids faced several hurdles. Although the strain of rice developed by Potrykus and Beyer produces a vastly increased amount of carotenoids, it remained well short of the amount that is needed. To meet the minimum daily requirement of vitamin A from a strain that they provided to rice experts in the Philippines in 2001, children would have to eat 20 bowls a day. Also critically important was that the strain was not one that is widely cultivated in Asia. Researchers worked initially with a strain that was known to be comparatively easy to transform (accept new genes). The strains of rice that sustain Asia (the "Indica" cultivars) are not as easy to engineer.

Fortunately, there has been steady progress. In 2003, the Indo-Swiss Collaboration in Biotechnology reported that it had engineered an Indica strain that, eaten in reasonable amounts, could provide 30% of the daily requirement of vitamin A. Meanwhile, even before news of events in China, public fear of GM rice was abating. For example, in 2005, authorities in India agreed to permit the cultivation of certain GM crops in some of its northern states. China is already a major grower of genetically engineered corn, soy, and cotton, so it should readily accept an improved strain of rice.

Given the debate over GM food, the work to create genetically engineered rice that is highly enriched in vitamin A raises the question of whether other solutions might be available. In addition to using standard hybridization techniques in developing rice strains, there is the obvious possibility that other plants might offer an easier solution. For example, Michael Dickson, who recently retired from the New York State Agricultural Experimental Station in Geneva, New York, spent 30 years studying a cauliflower plant with orange fruit discovered in a field in 1970. He finally developed a hybrid plant with very high levels of precursors to vitamin A (much more than in many green and leafy vegetables that are usually regarded as the best source). Of more relevance to the third world, scientists at the University of Newcastle recently reported their analysis of the carotenoid content in 12 varieties of bananas that had been chosen based on the yellow hue of their flesh (an indicator of carotenoids). Four of the 12 varieties have substantial levels of precursors. Further breeding efforts might create a banana so rich in carotenoids that, by eating two a day, a person would eliminate all risk of vitamin A malnutrition. Banana cultivation in desperately poor nations confronts a quite different set of obstacles (critically, a significantly greater need for land and a higher cost of cultivation per hectare) than does rice growing, but it is heartening that there are naturally available sources of vitamin A.

In April 2005, hopes to avert vitamin A blindness with genetically engineered rice took a dramatic leap forward. A team of scientists from Syngenta reported that they had created a new version of golden rice by replacing the daffodil gene for phytoene synthase with the equivalent gene from corn. The newly engineered rice accumulates levels of provitamin A that are more than 20-fold higher than in the earlier strain. Although the ultimate dietary yield depends on a myriad of factors, it appears that Syngenta has created a strain of rice that could, if consumed regularly by children in reasonable amounts, avert blindness. Syngenta has agreed that this rice will be freely available to the small-scale farmers in the third-world countries. Since grain can be distributed rapidly, it is possible that in just a few years we will begin to see a marked reduction in blindness due to vitamin A deficiency. Of course, the potential for dramatically reducing childhood blindness will depend more on economic and political factors than on scientific factors.

The problem of micronutrient deficiency encompasses far more than vitamin A. According to experts at the U.S. Plant, Soil and Nutrition Laboratory at Cornell University, more than one-half of the people on earth suffer from micronutrient malnourishment (sometimes called hidden hunger). In addition to vitamin A deficiency, the other common ailments are deficiencies in iron, zinc, selenium, iodine, and vitamin D. In some parts of the world, it is also not uncommon to find many people who are seriously deficient in vitamin C, vitamin B2 and B6, and folate. An outbreak of scurvy (vitamin C deficiency), a disease almost unknown in the western nations, recently occurred in Afghanistan. Public health experts claim that micronutrient deficiencies account for nearly two-thirds of all childhood deaths each year.

Some experts think that iron deficiency anemia is the most common disease in the world, affecting about 2 billion people. It is an insidious disorder to which children and young women (especially in pregnancy) are particularly liable. The main clinical signs are weakness and lethargy. People who are chronically deficient in iron have weakened immune systems, are ill often and for longer periods than others, and have a greater chance to die of infections. In a country with an adequate health care system, iron deficiency anemia is easy to diagnose and treat with supplemental iron. However, in much of Africa and large parts of Asia, the disease goes unrecognized. Even where it is recognized, affected persons often do not have access to supplemental iron or to food fortified with iron.

Just as with vitamin A deficiency, the best hope to greatly reduce the prevalence of iron deficiency anemia may be to use genetic engineering to create rice that is high in elemental iron. Potrykus and his colleagues in Zurich have made important advances toward that goal. In 2002, they announced that they had transferred the gene for ferritin from a bean plant into a strain of Japonica rice, effectively doubling its available iron. Other researchers, using a ferritin gene taken from soybeans, have tripled the iron availability. Eating just 8–10 ounces of such rice could supply about 20% of a pregnant woman's daily iron needs. This is not good enough, but it is a start.

Unlike the distribution of nutritional supplements or the fortification of foods, the delivery of iron through genetically engineered rice (or other plants) carries with it the potential for a permanent solution to the micronutrient famine in the third world. High-iron rice could be widely grown without disrupting local food economies. If plant breeders develop a Japonica strain that is both iron-rich and hardy, the improvement in human health could rank with that attributed to vaccines and antibiotics.

Each day in the world, about 10,000 children die a hunger-related death. Hard as it is to believe this number, it is almost certainly a conservative estimate. Many of these deaths are not due to famine. Too often the food is available somewhere in the country, just not available to the mouths of helpless children who need it. Using molecular genetics to develop hardier, more nutritious strains of rice will not solve the agonies of the third world, but it will likely help to save millions of lives.

PART
4

SOCIETY

Thomas Jefferson. (Courtesy of the Library of Congress.)

History

Popular historians and their readers love to unearth unanswerable little mysteries involving famous people long dead. History buffs seem inexhaustibly interested in the quirks, peccadillos, and, especially, sexual proclivities of famous artists and politicians. For example, a recent book on Abraham Lincoln, *The Intimate World of Abraham Lincoln* (by C.A. Tripps, who died shortly after its publication), is a painstakingly constructed argument that he was a homosexual. Even though most top Lincoln scholars dismiss the work as a web of speculation, it captivated readers. Martha Nell Smith, a professor of English, devoted years to the research supporting her claim that the poet Emily Dickinson was a lesbian who conducted a long relationship with her brother's wife. Musicologists have long speculated whether Franz Schubert was gay. Other favorite topics include whether or not famous men produced unacknowledged offspring, whether historical figures suffered from particular diseases, and whether the famous met their demise in ways other than those reported by the standard histories. One of the most visible recent such conjectures is that the great explorer, Meriwether Lewis, was murdered, a challenge to the consensus view that he shot himself.

Over the last two decades, advances in DNA analysis, especially the ability to recover and study ancient DNA, have provided popular historians and scholars alike with a new approach to answering questions that range from the trivial, gossipy sort to those that qualify as genuine (albeit usually minor) scholarship. DNA analysis has emerged as an important evidentiary tool. DNA typing has even been used to address some of the most enduring historical legends.

During the mid-19th century in the United States and France, it was widely believed that, contrary to the official records of the Committee on Public Safety (the revolutionary body that then ruled France), Louis XVII, the son of Louis XVI and Marie Antoinette, both of whom died under the guillotine during the French Revolution, did not die shortly thereafter in a

Paris prison. Although he was said to have died of natural causes on June 8, 1795, a rumor raced through Europe that the body of another child in the prison had been placed in his cell, and that royalists had aided the dauphin's escape to the United States. In 1814, the official historian of the restored monarchy fanned the fires when he declared that Louis XVII was alive (he provided no support for the claim). During the first half of the 19th century, hundreds of men claimed to be the dauphin, and a few others, the most famous of whom was John James Audubon, the naturalist (who was the right age and had been adopted without knowledge of his past), were rumored to be. Over the last two centuries, about 600 books have been written exploring the credibility of the dauphin story. To this day, descendants of the most convincing pretender, a man who grew up in Germany named Karl Wilhelm Naundorff, claim French royal lineage. One good measure of the immense popularity of the dauphin escape legend is Mark Twain's decision to write a character claiming that title into *Huckleberry Finn*.

One of the most intriguing threads of the dauphin legend is the claim that the physician who performed the autopsy on the dead boy in the prison stole his heart. During the first half of the 19th century, a desiccated organ alleged to be the boy's heart passed through several hands, eventually winding up in the possession of the Spanish wing of the Bourbon royal family. In 1975, it was returned to France and was interred in the royal crypt in the Saint Denis basilica, the resting spot of the dauphin's parents. In 1999, the French historian Phillippe Delorme obtained permission to subject the heart to DNA testing. Scientists at two different laboratories compared the results of mitochondrial DNA analysis with the results of similar studies of a lock of hair said to be that of Marie Antoinette and of tissue provided by two of her living descendants. The results unequivocally showed that all the individuals who were tested were related. These findings do not irrefutably establish that the heart came from the boy who died in prison or that Marie Antoinette was the mother of the child from whom the heart was taken. But they do provide powerful evidence that Louis XVII did in fact die in prison in 1795. A small group of die-hard French royalists were delighted with the scientific results. After wrangling consent from the French government, they held a ceremonial procession to rebury the heart of their dauphin in the Saint Denis basilica.

DNA and other forensic analysis of long-dead celebrities seems to fascinate Europeans. Among others, the great poet Petrarch and the Renaissance painter Giotto have been exhumed for testing. One research team has managed to find evidence of a cancer-causing mutation in the remains of a colon tumor that killed Ferdinand of Aragon, the king of Naples, in 1494! Recently, Gino Fornaciari, a paleopathologist at the University of Pisa, was given permission to exhume 49 members of the Medici family that once ruled Florence with an iron hand. Forensic studies might confirm or refute allegations that many of them died violently. DNA studies might determine whether infectious agents such as malaria ended their lives.

A somewhat less grisly example of the use of DNA testing to resolve a minor, but legitimate, question in American history is the quest to ascertain the birthplace and family history of Miles Standish. Although he was probably not a religious dissenter, Standish was the military leader of the 102 colonists who survived the 1620 voyage of the *Mayflower*. Contemporary diaries describe him as a small man with a fiery temper. On several occasions, especially near the town of Weymouth (in what is now Massachusetts) in 1623, he led the forces that saved the tiny colony from attack by Native Americans. As the Pilgrim foothold grew stronger, Standish took on important civilian duties; from 1644 to 1649, he was the treasurer. His military acumen apparently fell short when it came to wooing. His courtship problems were immortalized by the poet Henry Wadsworth Longfellow in 1858. Before his death in 1656, Standish founded the Massachusetts town of Duxbury, which is relevant to the contest over his ancestry.

There is no doubt that Miles Standish was English, but no one is certain when or where he was born. Standish is a relatively common English surname, and it would not be surprising if all men so named have a common ancestor. Currently, two major factions claim they are descendants of the line most closely related to the famous Pilgrim. The stronger claim appears to lie with the Lancaster branch of the Standish family. One reason is that the village of Chorley in northwest England was once the site of an estate called Duxbury Hall. There is also a Standish pew in the ancient village church, and several Standish family members who died in the 16th century lie in the crypt. The other branch, which includes a number of Americans, claims that Miles Standish was a Manxman. Old land records on the Isle of Man in the Irish Sea show that a family named Standish lived on a farm called Ellenbane. Miles Standish mentioned the Isle of Man in his will.

Several years ago, Caroline Kendall, who spent many years as the chief genealogist for the *Mayflower* descendants, decided to attempt to resolve the issue through DNA analysis. Her goal was to find American men who were descendants of the first Pilgrims and who could document continuous father-to-son lineage since 1620. Because the Y chromosome is transmitted vertically to sons, a man alive today has (with the rare exceptions caused by mutations) the same DNA sequence as that of his male ancestors hundreds of years ago. Kendall's plan was to compare the Y chromosome DNA information derived from the American descendants with the same material taken from men in the English families.

Unfortunately, there are very few men who can satisfactorily prove their lineage for such a long stretch. However, one man, a Benedictine monk named Benjamin Standish, who can document his family history back to 1780, has some evidence that his ancestors lived in Chorley. A comparison of his Y chromosome DNA with that of two Americans claiming descent from Miles Standish does not show a perfect match, but it is so close that it permits a strong inference that the three have a common ancestor. The year 2006 is the 350th anniversary of the death of Miles Standish. To clinch their claim to him as a favorite son, the residents of Chorley plan to open the Standish crypt in the hope of getting tissue for DNA testing to confirm their claim.

The most scandalous and vitriolic historical debate involving DNA analysis has been whether or not Thomas Jefferson sired children by Sally Hemings, one of his slaves. In December of 1998, Eugene Foster, a pathologist, published an article in *Nature* titled "Jefferson fathered slave's last child." Foster has collected DNA samples from known living descendants of Eston Hemings, youngest son of Sally, who lived at Monticello for more than 30 years. By performing Y chromosome analysis on the Hemings samples and on DNA obtained from acknowledged descendants of Thomas Jefferson, Foster was able to show that both lines had the same pattern. Based on DNA studies and historical records, they argued that the best explanation for this finding was that Thomas Jefferson was the father of Eston. An accompanying editorial by Eric Lander, one of the architects of the Human Genome Project, and Joseph Ellis, a Jefferson scholar with a revisionist bent, stopped just short of asserting that Thomas Jefferson was the father and that the case was closed.

The assertion that one of America's founding fathers conducted a long-term sexual relationship with a slave reverberated loudly throughout the United States. The scientific study and reactions to it were reported widely. The zenith may have been the day Oprah Winfrey united descendants from the Jefferson family and the Hemings family on her show. One key aspect of the investigation was regularly misrepresented. For example, *Time* magazine wrote that "Similar DNA tests, as the world now knows, established the youngest of Heming's sons, Eston, *was* Jefferson's child." Y chromosomal studies cannot support such a flat assertion. What the investigation did show was that Eston was the son of a man with the Y chromosome haplotype (pattern) characteristic of Thomas Jefferson *and of all other men who shared a common male ancestor with him.* During the period in which Sally Hemings conceived and gave birth to six children, there were 25 such men in Virginia, of whom 8 lived near and were regular visitors to Monticello. The correct characterization of Foster's finding was made in a report filed a year later by the Thomas Jefferson Foundation: A member of the Jefferson family fathered at least one of Sally's children *and* historical analysis concerning the Jefferson men living in Virginia at that time is most compatible with Thomas Jefferson's being the father. The latter claim is more difficult to prove than the former.

The Foster claim set off a minor academic war. Assertions that Jefferson had a sexual relationship with Sally (who was reported in contemporaneous documents to be of light complexion and a great beauty) were circulated as early as 1802, first as a political "dirty trick." Scholarship on the subject has waxed and waned over two centuries. Perhaps influenced by the rise of African-American studies in academia, during the last quarter-century there has been renewed interest in (and several books written about) the Jefferson–Hemings story. Although it at first appeared that the *Nature* paper (which appeared at the apogee of public fascination and faith in genetic evidence) would be the last word on the subject, the story has played out quite differently.

In mid-2001, a group of scholars convened by the newly formed Thomas Jefferson Heritage Society reported their findings. Their consensus was that Randolph, the younger brother of Thomas Jefferson, or one of Randolph's five sons, was more likely than our third President to have sired Eston. Among their many points were that Randolph and his sons (who lived only 20 miles from Monticello) had easy access to Sally, that they were known to have sexual relations with their slaves, that Jefferson was in his sixties when Eston was born, and that in the years after his return to Monticello after 8 years as President, Sally gave birth to no more children. Furthermore, there is no record that Sally (whom Jefferson freed) or any of her children ever alleged that Thomas Jefferson was the father. Ironically, DNA studies did show that another child of Sally Hemings, Thomas Woodson, who was by oral history linked with the President, was definitely not his child.

In 2005, the majority view is still that Thomas Jefferson did father children by Sally Hemings, but it is not universally held. The evidence currently being used to attribute Eston's paternity to Thomas Jefferson would probably not support a modern paternity suit. Why? The results of DNA testing of a mother–child–alleged father trio can absolutely exclude a putative father. However, it is not possible to obtain results that absolutely prove fatherhood. The Y chromosome studies demonstrate only that a member of the Jefferson family fathered Eston. The record does not demonstrate that Sally ever identified Thomas Jefferson as the father. Without the woman's accusation, Y chromosome evidence alone is insufficient. Perhaps a better test is to consider the debate as analogous to a lawsuit brought to assert an inheritance right by a man claiming to be the unacknowledged son of the deceased. Construing the DNA test results and the historical record, a probate judge probably has sufficient evidence upon which to rule that a descendant of Eston can claim a blood relationship with Thomas Jefferson. However, absent the discovery of new historical evidence (contemporaneous letters, diaries, and the like), I doubt the level of intimacy in the Jefferson–Hemings relationship will ever be finally resolved.

Sadly, seven years after the DNA study that linked the Jefferson and Hemings families, the two groups have grown far apart. In 2004, when 20 Hemings descendants sought to attend the annual Jefferson family reunion (which drew about 80 people) they were told that they would only be admitted if escorted by a (white) Jefferson family member. The Monticello Association has excluded Hemings descendants from membership and prohibited any of its members from inviting more than two Hemings descendants to be their guests at the meeting. Apparently unflappable, the Hemings group is planning its own reunion at (where else?) Monticello.

PLAGUES

One of the most powerful and least well understood forces that has shaped the course of human history over the last few thousand years is disease. Sometime in high school or college, students read a paragraph or two about the Black Plague, the pandemic that swept in waves through Europe and Asia in the 14th century, killing about one-third of the population. But virtually no one learns that, during the 18th and 19th centuries, the rapid growth of cities and trade provided new routes for cholera, which had been endemic in India for centuries, to circle the globe. During that era, cholera found tens of millions of new victims. Syphilis, which by the early 20th century usually caused a chronic, progressive disease that killed slowly (often by eroding the aorta) was in the 15th century much more likely to be quickly lethal. The bacterium, *Treponema pallidum*, which causes syphilis, has coevolved with humans for several hundred years. Any self-respecting germ knows that it is better not to kill your host. Smallpox was endemic in Europe for so long that a combination of genetic changes in both host and pathogen markedly reduced its death rate. However, when it arrived in the New World, it killed far more Native Americans and Pacific Islanders than did any invading army or navy. Although cholera is rarely diagnosed in the United States today (only in a handful of immigrants), on three separate occasions (1832, 1849, and 1866), epidemics terrorized America, nearly paralyzing urban life in those summers. Today, the HIV/AIDS epidemic threatens the demise of human society in sub-Saharan Africa.

Over the last few years, historians have combined forces with molecular biologists to try to understand the origin and spread of infectious diseases. One of the first major successes was the demonstration in 1994 that an ancient Peruvian mummy had died of tuberculosis, thus refuting the long-held theory that this disease was brought to the New World by Europeans. Such investigations are not the enthusiasms of dilettantes. Late in 2004, the World Health Organization speculated that if its worst scenario unfolded, a severe pandemic of Asian bird flu could quickly kill as many as 50 million people. The Centers for Disease Control and Prevention in Atlanta currently cling to a low estimate of only 2 to 7 million deaths, while other experts assert the toll could exceed 70 million! The most feared strain of the virus, called H5N1, killed only 32 humans in 2004. But H5N1 has a wide range of animal hosts (including ducks, mice, and cats) and has killed a high proportion of the few humans who have been infected. Should H5N1 mutate to make it easier to infect humans, the result could be devastating. In addition to providing historians with more knowledge about where and when they arose and the magnitude of their impact, DNA analysis of the past plagues might provide crucial insights into how to fight others in the future.

In the mid-19th century, more than 1 million people in Ireland died from starvation when a fungus-like disease devastated the potato crop year after year. While the brutal disregard of the English landowners to the plight of their starving peasants greatly exacerbated the death toll, the havoc wrought by this plant pathogen was the single most important factor in the mass emigration to the United States. Over the last century, scientists have demonstrated that the Irish famine and many other crop failures were caused by an oomycete called *Phytophthora infestans*, currently regarded as world agriculture's most destructive pest!

For decades, experts thought that a particular strain of *Phytophthora infestans* called *1b* caused the Irish famine. Recently, a team led by Jean Ristaino, a plant pathologist at North Carolina State University, completed DNA analysis of samples taken from 150-year-old potato plant leaves (dating from as early as 1845) that it tracked down in botanical collections from around the world. Using DNA analysis, the scientists found that the Irish potato famine was almost definitely caused by a different, less common, strain called *1a*. Perhaps their most interesting discovery was that strain *1a* is native to South America. Ristaino's work suggests that the pathogen that killed 1 million Irish arose in the Andes mountains in South America (which is also the region in which people first cultivated potatoes) and traveled with potatoes that were sent to Europe in trade.

The Irish potato famine pales in significance to the Spanish influenza pandemic of 1918–1919. Historians agree that the virus killed 20–30 million people (far more than died as a direct result of battle during the First World War) as it raced around the globe. By studying mortality data week by week in 45 major cities, epidemiologists have concluded that the flu of 1918 did not spread more easily than other influenzas. Its two distinguishing features are a case fatality rate ten times greater than in similar epidemics, and that the healthiest individuals were the most likely to die. For example, the 1918 flu took the lives of tens of thousands of young men in the trenches of World War I. Most victims died rapidly of hemorrhagic pneumonia, usually within two weeks of becoming ill. This is dramatically opposite to the usual pattern in which influenza claims the elderly, infants, and the debilitated.

Armed with the tools of molecular biology, during the 1990s, virologists realized that if they could obtain tissue which still harbored the 1918 influenza virus, they might be able to learn why the virus was so deadly. A search of death records in Norway revealed that seven men who had died of the influenza in the remote village of Longyearbyen (which is near the arctic circle) had been rapidly interred in permafrost. Aided by the use of ground-penetrating radar, in 1999, scientists were able to exhume the permanently frozen bodies and obtain lung tissue from them.

DNA analysis of the Spanish flu virus samples found in tissue taken from the exhumed bodies suggests that a mutation in a gene that codes for a protein called hemagglutinin made it easier for the virus to enter human cells. Some experts now think that the Spanish influenza of 1918 was a bird flu that jumped species shortly before the pandemic began, but the assertion is hard to prove. Many scientists are convinced that the change in hemagglutinin caused the virus to provoke an unusually violent inflammatory response in the host. Ironically, the healthiest hosts might have mounted too strong a defense, one in which their white cells released chemicals that severely damaged their blood vessel walls as they were destroying the virus. Many of the victims of the Spanish flu may have died from "friendly fire." If this theory is correct, it may mean that if, or rather when, humanity is next engulfed in a pandemic, the administration of anti-inflammatory drugs might help reduce the death rate.

ANIMAL HISTORIES

Just as in efforts to decode the history of hominids, DNA analysis is being used to answer questions concerning the history of animal life. For example, it is being used to reexamine the standard teaching that, coincident with the arrival of humans from Asia, there was a great die-off of mammoths, short-faced bears, Beringian (the word refers to the large landmass that once connected Alaska to Siberia) bison, and other large mammals in North America. For decades, professors taught that the main cause was human predation. I have always doubted this hypothesis; in my view, the continent is much too big and the number of humans then was much too small to kill off such formidable species so quickly.

In 2004, a large group of scientists, led by Alan Cooper, an evolutionary biologist at Oxford, built a convincing case against the human predation theory. They analyzed a 685-base pair sequence of mitochondrial DNA extracted from more than 400 bison fossils and used radiocarbon techniques to date more than 200 of them. Their analysis showed that all the animals shared a common ancestor that lived about 140,000 years ago, that by 100,000 years ago, bison had spread as far south as present-day Mexico, and that beginning about 37,000 years ago, the bison population went into sharp decline. Their analysis also suggests that the last ice age (22,000 years ago) cut off contact between the northern and southern herds, and that, starting about 8,000 years ago, changing habitats (mainly new forests) isolated small populations for millennia. For those who hew to the Cooper group's argument, it would follow that humans delivered the coup de grace to bison (and, perhaps, other large mammals), but that the population of bison had already been sharply reduced by climatic change. Incidentally, the fossil record shows that over the last 5,000 years, the North America bison population rebounded, only to crash again to the edge of extinction at the hands of sport hunters in the mid-19th century.

Alan Cooper and his colleague, Beth Shapiro, have also undertaken fascinating studies to better understand the sad history of the dodo, a poster child for extinction. The dodo (*Raphus cucullatus*) was an ungainly bird with large strong legs, tiny wings, a short neck, and an enormous, thick beak. An adult weighed more than 40 pounds. When Portuguese colonists arrived in Mauritius in the late 1500s, the dodo, which had become flightless during millennia of life on an island blissfully free of predators, was—to coin an apposite term—a sitting duck. A population that almost certainly was in the millions became a favorite entree for man as well as some of his companion animals. Newly arrived rodents especially enjoyed dodo eggs. The last certain sighting of a dodo was 1662, but some scientists argue that a few may have lingered until 1690.

All that remains of the dodo, immortalized by Lewis Carroll in *Alice in Wonderland*, are a few partial skeletons scattered among museums of natural history, and part of a desiccated head and foot at the Oxford University Museum. The availability of even a little preserved soft tissue has led Cooper and others to embark on a quixotic journey. The goal is to recreate the dodo. To realize this goal, scientists would have to assemble most of its DNA fragments and then use cloning and other technologies to make a fair copy. To do this, it is crucial to determine the dodo's closest living relative. It turns out that the dodo is really just an unusual pigeon that, like many other birds, underwent rapid evolutionary change while living for millennia on islands. Its nearest living relative is the Nicobar pigeon (*Caloenas nicobarica*), which lives on eponymous islands in Southeast Asia. Cooper and Shapiro hope that someday a Nicobar pigeon will be a surrogate mother to a cloned dodo. Given the rapid advances in genomics, it is difficult to reject the possibility that a dodo cloning project might succeed, but it is a real long shot. I think it unlikely that scientists would be able to obtain funds to do the massive amount of work needed to assemble the dodo DNA fragments (if, indeed, they could all be recovered).

Horse-racing fans will be intrigued to learn that geneticists have used DNA testing and pedigree analysis to determine the relative contribution of important founder horses to the current population of Thoroughbreds. About 1 million horses (half of them alive now) have been registered in the

Stud Book (the official record on Thoroughbreds), a recognition reserved only for horses born to parents that are also both registrants. The modern population traces its origin to a relatively few horses that were brought to England about 300 years ago. DNA studies have demonstrated that the founding animals came mostly from Saudi Arabia, Turkey, and the Barbary Coast. Such studies have also shown that certain animals have contributed a vastly disproportionate amount to the Thoroughbred gene pool. Among stallions, a horse called Darley Arabian (born in 1700) has been shown to be the source of 95% of all the Y chromosomes in the world Thoroughbred population. Essentially, most Thoroughbred males can claim him as an ancestor. Among mares, one is said to be the source of more than 15% of all the mitochondrial DNA. It appears that as few as 30 founder horses are responsible for 80% of the modern gene pool. The good news is that they were a genetically diverse group.

Great racehorses often earn much more as studs than they did on the track. The most valuable horse in the world, Kentucky Derby-winning stallion, Fusaichi Pegasus, was sold to stud for $64 million. His current stud fee is $150,000 per insemination. In season (mares only ovulate in spring and summer), he works three times a day. Needless to say, Fusaichi Pegasus is heavily insured.

But how much do genes contribute to the making of a great racehorse? The world record for the mile and a quarter was set by Secretariat more than 30 years ago. In horse chronicles, that is about 10 generations (of highly controlled breeding) ago. Why have none of his progeny broken his record? One purely environmental reason is that modern tracks are maintained to be somewhat softer (and, therefore, slower) on the animals' hooves. But as we learn more about the horse genome, we are discovering genes that influence heart size, oxygen-carrying capacity, temperament, and a variety of other traits that intuitively seem to be important in building a great racehorse.

Will DNA testing take horse breeding to a new level? In the long run, I think so. For example, some experts think that horse breeders have relied too much on the bloodlines of stallions and have not adequately valued the genetic heritage of mares. DNA testing may alter that. Under any scenario, environmental factors will also continue to be hugely important. Just as with human athletes, nutrition and training, especially approaches that rely heavily on modern sciences, may be more important than a quixotic search for just the right combination of genes.

Major Robert Harry Schuler. (Photo courtesy of Laurie Boone.)

DNA Forensics

The remains of Major Robert H. Shuler Jr., a distinguished Air Force pilot who died in combat, have come home. Major Shuler did not die in Afghanistan or Iraq. He died some 40 years ago in North Vietnam. It was October 15, 1965. Shuler, a squadron leader, was lingering in enemy airspace to see whether a fellow pilot had safely ejected from a damaged plane. That is the last that anyone saw him alive. Officially, Shuler was listed as "missing in action." As the years passed with no word of his fate, his wife remarried; her new husband adopted her two daughters. Shuler's parents died in the late 1970s; his daughters are now middle-aged women.

In 1999–2001, reacting to the development of new methods to analyze DNA, the Department of Defense, working with the government of Vietnam, renewed its efforts to identify remains found near battlefields of the Vietnamese war that might be those of U.S. servicemen. Forensic scientists were able to construct DNA profiles on hundreds of remains, but that by no means ensured identification. During the 1960s, DNA identification tools did not yet exist, so there was no database against which to run DNA profiles of anonymous remains. In the poignant closure to Major Shuler's case, the dedication of another Vietnam veteran and an old shaving kit were as important as the new science.

When Dennis L. Wolfe, the veteran, who like Shuler was a native of the region around Elmira, New York, heard that Shuler was after 40 years still missing in action, Wolfe named the local chapter of the Vietnam Veterans of America in his honor. Fred Shuler, a nephew of the major, found Wolfe when he was doing research on his uncle. It was Wolfe who put Fred in touch with the military scientists who were seeking any object that might have traces of his dad's DNA. Under the laws of genetics, Fred's dad—Major Shuler's brother—carried one-half the genes that his brother, the major, inherited. In addition, both men inherited the same mitochondrial DNA from their mother (via the cytoplasm in her eggs). Nuclear DNA does not last long in the hot, wet soil of the jungles of Southeast Asia, but mitochondrial DNA (a small circle of DNA of which each cell has hundreds of copies), which can be extracted from teeth and bones, sometimes does. The forensic specialists were able to use mitochondrial analysis to construct a genetic profile of newly recovered human remains, which they entered into a database. Next, they needed tissue samples from relatives of men long missing in action to determine whether the DNA profiles were similar enough to establish the identity of the remains. Stored away in the basement of Fred Shuler's home was an old shaving kit that had belonged to his father. Forensic experts tested hairs from Fred Shuler's brother, a man who had been dead for a decade. They found a match. The hair DNA had a profile that would be expected of a brother of the man whose remains were thought to be those of Major Shuler.

At this writing, this story is not yet over. Major Shuler's next of kin are his two daughters, who lost him when they were very young. Now middle-aged, they must have often wondered what happened to him, but they probably never thought that they would ever attend his funeral. Yet, against all odds, someday, probably before this book is published, they will attend a funeral with full military honors for Major Shuler as he is laid to rest in Elmira, New York.

The story of Major Shuler is just one of thousands of poignant, yet fascinating, stories that have grown out of advances in molecular genetics. As we have seen, the new DNA identification tools are both extraordinarily sensitive and extraordinarily accurate. In addition to greatly assisting in the resolution of painfully difficult mysteries of identity, they are revolutionizing criminal law. Indeed, it may be that DNA evidence has had greater impact on the resolution of crimes than has any other technological development in human history. The ever-growing impact of DNA evidence on the law started barely 20 years ago.

The revolution—at first called DNA fingerprinting—was discovered in the fall of 1984 in the laboratory of a 34-year-old geneticist named Alec Jeffreys, who conducts his research at Leicester University in the midlands of England. As is often the case in basic research, the discovery was serendipitous. Jeffreys was mainly interested in using molecular techniques to study how genes evolve (one of his studies focused on sparrows). But his team was also looking into a side project, trying to understand the nature of repetitive sequences of DNA in the genes that codes for human myoglobin, a muscle protein. They hoped their studies might lead to better mapping techniques—tools for figuring out where specific genes are located within the genome. With the hindsight of 20 years and since the completion of the Human Genome Project, this study might seem quaint, but in 1984, we knew the locations of very few genes.

In late September, the team led by Jeffreys discovered that the human genome contained many regions of DNA that were hypervariable—that is, there were short repetitive segments that varied substantially among individuals. They could demonstrate this because the variability manifested as differences in the number of copies of the repeating unit of DNA. The differing size units could be separated by forcing the samples to migrate in a gel under the influence of an electrical current. If one person has nine copies of a repeating unit and another has seven, that difference will manifest as a difference in position on an X-ray film taken of the DNA samples after they have been labeled with a radioactive probe and forced to migrate in a gel under the influence of an electrical current. Just as a bar code is used to denote specific items in the grocery store, the X-ray pattern is a unique identifier. With the sole exception of identical twins, assuming that one uses probes to measure a sufficient number of variable DNA repeats, one can construct a pattern on every person which would be unique to that individual among the 6 billion people on the planet.

The validity of this bold statement is crucial to the revolutionary impact of DNA science on forensics. Imagine an identification system that involves 10 different regions of the genome (ideally, located on 10 different chromosomes), each of which has a high degree of variability. Let us further assume that the chance of any two unrelated people having the same copy number for any one of the 10 regions is only 10% (which is roughly the case for many regions). If so, the odds of any two persons selected randomly from any population on the planet having exactly the same set of DNA fingerprints is only 1 in 10 billion (raise 1/10 to the 10th power). If this is so, then one could test every person on the planet and there would be only about 1 chance in 2 that even one result would match that of the first person tested!

Professor Jeffreys and his team readily anticipated that genetic fingerprinting could be a boon to resolving crimes of violence such as rape and murder, especially if the scientific method worked on old crime scene samples. They contacted the Home Office (akin to the FBI) and were loaned crime scene specimens that were several years old on which they were easily able to show that they could obtain a DNA profile. Jeffreys' discovery came just as the second of two horrific rape murders of teenage girls occurred in a nearby village. A year or so later, frustrated by an unsuccessful investigation, which was complicated by the confession of a mildly retarded teenage boy that the police concluded was untrue, the Narborough police inspectors asked Jeffreys for help. After analyzing the crime scene samples, he was quickly able to confirm that the young man who confessed had in fact not committed the crimes. In addition, Jeffreys was able to confirm (through DNA analysis of semen stains) that the two victims were both killed by the same man.

The murder investigation continued to take a bizarre path. The police asked hundreds of men in the villages where the murders occurred to voluntarily agree to undergo DNA profiling. None of the profiles matched that of the killers, but, oddly, two profiles, ostensibly from two different men, were the same. Such a finding contradicts the essential premise on which DNA testing is built—that all persons save for identical twins will have a unique genetic profile (there were no identical twin adults living in the area). How could that be? The answer provided the clue to the capture of the killer. On questioning, the man whose profile appeared in duplicate confessed that, in addition to his own sample, he had also stepped in for a friend, who had said he was afraid of being stuck by a needle and always fainted. The second man was rapidly found and tested. His DNA matched that of the DNA in the semen found in the two dead girls.

Over the ensuing five years (1987–1992), DNA evidence became the hottest topic in criminal law enforcement. Although the United Kingdom maintained the lead, use of the science spread quickly in the United States. Time and again, judges permitted prosecutors to introduce what appeared to be extremely persuasive DNA evidence over the sputtering objection of defense attorneys, caught off guard. However, by the early 1990s, the U.S. criminal defense bar, aided by a small, but vocal, group of scientific critics, began to raise judicial doubts.

The crucial issue in the defense counterattack was the scientific strength of the claim being made by the experts testifying for the prosecution that the odds that a randomly selected person would have a DNA profile that also matched that of the crime scene sample were on the order of 1 in 1 billion or less. At bottom, this claim is based on the assumption that each DNA probe used in the identification system is independent of every other. But what proof supported that? The critics sometimes successfully argued that there were an insufficient number of studies of the DNA markers among various populations to demonstrate the truth of the so-called "product rule" (which allows the odds on matching at each marker to be multiplied together to reach a final odds of matching).

Debates over the statistics roiled hundreds of trials in the early 1990s. In 1992, a report on the proper uses of DNA evidence commissioned from the neutral National Research Council refused to endorse the product rule, fanning arguments in the courtroom. Three years later, a second NRC study and a large number of studies in the field of population genetics did successfully validate the product rule, which is now routinely admitted into testimony in the United States and Europe.

Without doubt, the watershed event in the history of DNA forensics is the celebrated murder trial of the Hall of Fame Buffalo Bills running back, O.J. Simpson. On June 12, 1994, Nicolle Brown Simpson (who had divorced him in 1992, alleging repeated physical abuse) and her companion, Ronald Goldman, were found dead outside Brown's condominium in Los Angeles. Five days later, Simpson, the prime suspect, set the tone for the ensuing trial when he deviated from his plan to turn himself into police and went on a two-hour car chase that was ultimately watched live on TV by 100 million people.

In the ensuing eight-month trial, one of the most thoroughly reported in U.S. history, the validity of DNA evidence derived from analyzing blood on several items of clothing, including a glove that allegedly belonged to O.J., was a key topic. Among the 150 witnesses who testified were leading population geneticists and DNA forensic scientists who had analyzed the bloodstains. Although the lab results unequivocally found that some of the bloodstains derived from O.J., the defense was able to overcome that testimony by raising the possibility that the evidence had been "planted" by a racist police detective. One of the most famous moments of the trial (devastating to the prosecution case) was when O.J. was asked to put on a bloodstained glove that turned out to be too small (probably due to shrinkage by the blood), leading defense attorney, Johnnie Cochran, at his closing argument to quip, "If the glove does not fit, you must acquit..."

Despite the acquittal, the daily reporting on the trial taught more Americans about the power of DNA evidence than could have been accomplished by a legion of professors. The "not guilty" verdict did not at all slow the use of DNA evidence in the courtroom. In the ensuing decade (1995–2004), thousands of criminal and civil cases have been resolved (often by plea bargaining or settlement) after a review of the results of DNA testing. For example, litigation to establish paternity, once common in American courtrooms, has disappeared. If an accused man is the father, the results of the DNA evidence are too powerful to counter. In 2005, one can confidently say that DNA identification evidence is now routinely accepted in legal matters. This recognition has generated two important new challenges for the criminal justice system: the DNA analysis of old evidence to reopen closed cases and the proper use of DNA felon data banks.

As the defense bar became familiar with DNA evidence, lawyers quickly realized that they had access to a powerful new tool with which to challenge the operation of the criminal justice system. In 1992, Barry Sheck, who would later become the "DNA" lawyer on O.J. Simpson's defense team, and Peter Neufeld started a legal clinic at Benjamin Cardozo Law School in New York City called the Innocence Project. They began to investigate cases in which men who had been convicted of rape and murder steadfastly maintained their innocence and in which there was crime scene evidence that, if subjected to DNA analysis, could potentially exonerate them. Many of the early cases involved men who had been given lengthy sentences for rape based on eyewitness testimony in the years before DNA testing became available.

The Innocence Project faced a difficult challenge. In our criminal justice system, once an individual has been convicted, exhausted his appeals, and gone to prison, the system regards the case as closed. As time passes, there is a strong bias against reopening it. Furthermore, in many states, there is not a clear procedure for doing so. However, most district attorneys will be quick to acknowledge that obtaining a conviction on an innocent man is one of their worst nightmares. Sheck, who had schooled himself in DNA science and won some key scientific allies, was indefatigable. In some cases, a judge who was reluctant to reconsider did so because Sheck convinced the prosecuting attorney to side with him on the petition.

During the last decade, the Innocence Project has had a remarkable impact. As of 2008, more than 200 men who were serving lengthy sentences for felony convictions have been exonerated and released because an analysis of DNA evidence from the crime scene demonstrated unequivocally that the convicted individual could not be the source of the semen or blood that was analyzed. The success of the Cardozo legal clinic has spawned many similar efforts in other law schools. The growing number of exonerations has called into question the validity of eyewitness testimony and the manner in which police organize and use "lineups." It also has pointed out the need, when there is biological evidence that could exonerate but that has not been tested, to establish a method for doing so. Over the past few years, at least 35 states have enacted laws to permit post-conviction DNA testing. Some, however, set out extremely limited circumstances under which this might occur. The Innocence Project has drafted a model statute to cover this issue and continues to lobby for a more liberal use of DNA evidence.

As the battle over the interpretation of DNA evidence was played out in the courtrooms of America from 1987 to 1995, the FBI was quickly moving forward to develop a platform to support a national network of state-based DNA felon data banks. These are computerized files made with a commonly used set of DNA probes that record the genetic identity of convicted felons before they are paroled. The idea of such a data bank is a logical extension of the century-old practice of fingerprinting, but (as I shall discuss) DNA profiling raises a number of difficult issues that do not complicate fingerprinting. California was the first state to collect tissue on convicted felons. Beginning in 1983, it began testing convicted sex offenders for protein markers found in human saliva that were also present in semen. The idea, of course, was to assist the state in solving future crimes, a good idea given the high frequency with which such felons strike again.

Shortly after Alec Jeffreys grasped the possible utility of DNA forensic evidence, law enforcement officers at the Home Office in the United Kingdom and the FBI in the United States quietly began to create the foundations for what are rapidly becoming national criminal DNA databases. From the start, officials in the United Kingdom were able to progress more quickly, largely because of the major difference in the governmental structures of the two nations. Police policy in the United Kingdom is largely set at the national level; in the United States, each state operates its own criminal justice system. In 1994, Parliament enacted the Criminal Justice and Public Order Act to amend a 1984 law on evidence to expressly permit the police to take samples for DNA profiling. In contrast, the FBI faced the challenge of developing a DNA test system that would be adopted by all 50 states, of assisting law enforcement in each state to obtain enabling legislation and adequate funding to participate, and of developing a secure network for data to be fed into a common bank for subsequent analysis in the investigation of future crimes.

Not surprisingly, state and local police were eager to participate. In the period from 1989 through 1999, all 50 states enacted laws that created DNA felon databases, a rare legislative event. Although some states did not provide adequate funding, most new programs were funded well enough to get at least one local DNA forensics lab up and running. Initially, the state laws applied only to persons who had been convicted of violent crimes such as murder and rape. However, as DNA technology became ever more sophisticated, lawmakers reacted by steadily expanding the scope of crimes to which the laws applied. Once it became clear that forensic scientists could use amplification techniques to isolate and study minuscule amounts of DNA left on the steering wheel of a stolen car or inside the glove that a burglar dropped at a crime scene, DNA felon laws were amended to cover car theft and burglary. In 2001, the British Parliament enacted a law (The Criminal Justice and Police Act) that permits the police to retain DNA samples on persons who were arrested but not prosecuted, and even on those who were acquitted!

As of 2007, the United Kingdom National DNA Database (a nation of about 68 million) held DNA profiles on over 4,000,000 people, making it on a per capita basis the largest database of its kind in the world. The bank, which is growing by more than 30,000 samples a month, is composed of samples recovered from crime scenes and from suspects. Since a new Criminal Justice Act came into force in 2004, any person arrested in England or Wales for all but the most minor offenses, will have his or her DNA sample stored for 100 years. Between 2005 and 2006, more than 45,000 samples taken from suspects or the crime scene were linked to profiles in the database. Among them were matches relating to 422 homicides and 645 forcible rapes.

The UK DNA Database now contains samples from more than 5% of the national population. The vast majority of samples have been collected from men. Currently, samples taken from about 10% of white men and 37% of black men in the United Kingdom are in the database, a fact that has led the Black Police Officers' Association to voice concern. Also troubling is that more than 24,000 samples have been collected from children or teenagers who have never been charged with a crime. In late 2004, a lawsuit was brought to challenge the policy of retaining samples from persons who were charged with an offense, but not convicted or arrested, and never charged. The Court of Appeal upheld the law, and an appeal has been filed with the European Court of Human Rights.

Currently, officials in the United Kingdom estimate that the probability of identifying one or more suspects by running a DNA profile of a crime scene sample against the national database is over 40%. As the database has grown, so have its successes. For domestic burglary, the identification rate when a DNA sample is recovered is over 40%, compared to only 14% when no usable sample is found. The Home Office estimates that about 50% of the samples at crime scenes that are analyzed help secure a conviction, with half of those resulting in a custodial sentence. They further estimate that each such sentence results in about eight fewer crimes being committed in the nation!

The development of DNA felon databases in the United States has been quite similar to the trajectory in the United Kingdom. As of 1999, every state had enacted the legislation necessary for it to participate in the national data-sharing system (called CODIS, which stands for Combined DNA Index System) developed by the FBI to enable investigators to search offender profiles created in the 50 different jurisdictions. As of mid 2007, more than 4.5 million samples had been collected and profiles. Officials at CODIS report that, on more than 50,000 occasions, they have found a match of new crime scene evidence with a profile in the database.

Some states now have highly developed felon data banking systems, and they are having an extraordinary impact. For more than a decade, the Bureau of Forensics in Virginia has been a leader in developing a state-based DNA criminal database and using it to resolve crimes. As of July 31, 2004, the Virginia database contained 221,242 samples. Unlike most other states that target convicted felons about to be paroled, since 2003, Virginia law also permits investigators to collect and store a DNA profile on any person *arrested* on suspicion of having committed a violent felony. Analysis of the impact of the database clearly shows that as it grows larger (collects DNA profiles on an ever greater number of convicted criminals), the chances that DNA evidence found at a crime scene for which there is no suspect will match a profile of a prior offender grows dramatically. These so-called "cold hits" have been steadily on the rise in Virginia; in August 2003 alone, evidence from 94 new crime scenes matched profiles in the extant database. In the first 18 months after the creation of the arrestee database, a total of 103 matches of crime scene DNA profiles to arrestee profiles have been made. Of these, 24 involved violent sexual assaults. About 80% of the matches would not have occurred if the database had been limited only to violent (rapists and murderers) offenders. About 37% of the violent crimes that have been solved because of DNA matches were committed by men whose only prior convictions were for crimes against property!

Although it is impossible to claim a cause-and-effect relationship, during the last decade (1993–2003)—the period in which DNA criminal databases have been maturing—the rape rate (defined as victimization rate per 1,000 persons age 12 and over) has dropped by over 60% and is historically at an all-time low. The broader violent crime index has witnessed a similar pattern.

The proliferation of DNA criminal databases has its critics. Not unexpectedly, the key theme has been privacy. Initially, those opposed to DNA criminal databases repeatedly challenged the enabling statutes in state and federal courts as a violation of the constitutional safeguard against unreasonable search and seizure provided by the Fourth Amendment. By 2004, more than a dozen such cases had been decided, all in favor of the state's right to create such databases.

Despite its great contribution to the criminal justice system, even wider use of DNA forensics is hampered by all-too-regular reports of sloppy or fraudulent work by scientists and technicians who are performing the analyses that will be crucial to the case offered by the prosecution. Although many important steps have been taken to reduce the risks of such events, the huge caseload backlog has created time pressures on many labs that increase the risk for errors. The best way to address this problem is by better funding of the labs.

Currently, the policy debate focuses on the permissible scope and use of the databases. Civil libertarians argue for restricting them to samples taken from felons serving time for violent acts against persons. Law enforcement officials have consistently favored much more expanded use. The reach of the statutes varies. Some state laws apply to all felons; others include arrestees. Recently, the FBI argued for the inclusion of juvenile offenders in the databases. As evidenced by Virginia, the trend is definitely to broaden the application of the databases. This has led some persons to assert that there should be a universal DNA database that includes a DNA profile on all citizens. They argue that this is the only way to avoid tainting the data banks with the racial prejudice that they think is rife in the criminal justice system. For both economic and political reasons, this is not likely to occur—unless, of course, fear of terrorism drives society to call for it. If someday our society adopts a universal DNA database, we will be assured of linking nearly every crime scene sample with an individual, potentially a huge boon to crime prevention.

One fascinating question in the policy debate is whether law enforcement officials should be allowed to conduct "low stringency searches." It has long been recognized that one of the single best predictors of whether a person will commit a crime is whether or not a close relative (especially the father) has already done so. In states with large DNA databases, it is relatively common that a DNA sample found at a crime scene does not match precisely any profile in the database, but does partially match with one in a way that indicates that the perpetrator of the new crime is a sibling, parent, or child of a person in the database. Databases of close relatives are routinely used to resolve identity in mass disasters such as the attacks on the World Trade Center. I think it likely that if the courts are asked to rule on the constitutionality of using partial matches to apprehend persons, they will approve them.

In my view, the great debate about the role of DNA evidence in law enforcement lies ahead. Two or three decades hence, the field of behavioral genetics may have matured to the point where district attorneys, parole boards, and others will become intrigued with the possibility of using DNA information to assist them in their work. For example, should the decision to parole a man who served time for vehicular manslaughter while he was inebriated be influenced by DNA tests that show he was born with a gene variant that confers a high proclivity to alcoholism? Since parole boards use a wide variety of evidence to reach their decisions, there is no fundamental reason that genetic test results should be excluded. Our society might decide that it cannot tolerate such deterministic evidence even if the underlying scientific claims are strong, but I think it more likely that we will embrace such information.

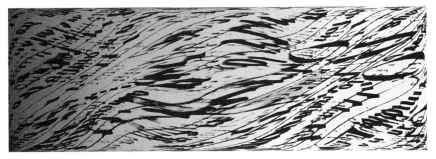

Frank Gillette, The Broken Code (for Luria), *2002. Chromographic print on aluminum. (Courtesy of the artist and Universal Concepts Unlimited, New York.) (Reprinted, with permission from Anker and Nelkin 2004.)*

18

Art and Language

When did the concept of heredity first take hold in the mind? When did people first begin to wonder whether mysterious biological forces set boundaries on their lives? How deeply can we trace the roots of genetic determinism? How can we measure the influence of advances in genetics on the zeitgeist of today? Humans have engaged in recognizably artistic (and, therefore, probably linguistic) activity for about 40,000 years. Surely, the minds that could conceive and execute the magnificent cave paintings in Lascaux grasped that children looked more like their parents than like others in the social group. How long ago did a grandmother for the first time brag that an infant looked like her? I would guess at least 20,000 years ago, but I would not argue with those who asserted that the better guess was double that.

We know little about what early civilizations thought about heredity. The rulers of ancient Egypt practiced incest (because the pharaohs were gods, only sisters or other close relatives were thought to be fit partners), and there is some archaeological evidence (skeletal abnormalities) of the biological price they paid for inbreeding. But the negative consequences of incest were apparently not sufficiently obvious to change culturally important traditions. Five millennia ago, the Egyptians worshipped gods that we regard as monsters. Osiris was incarnate as Apis, a bull with a man's body. Thoth, the god of knowledge, had the head of an ibis and the body of a man. The legendary sphinx, built by Chephren during the Fourth Dynasty about 4,500 years ago, has the body of a lion and the head of a man. The chimeric bodies of the Egyptian gods symbolized their ability to exist in both the earthly life and the afterlife. We will probably never know whether or not these unusual forms were influenced by notions of heredity.

Ancient Greece teemed with centaurs, fauns, gorgons, satyrs, giants, and other fantastic creatures. When Hera, the wife of Zeus, became jealous of his mistress, Io, she ordered the 100-eyed Argus to spy on her wayward husband. In *The Odyssey*, Homer sings of Ulysses's battle with the Cyclops. The oldest references to the unicorn are Greek, but this lovely creature also gallops through the folklore of ancient Persia, a fact that suggests it was imagined independently in several cultures. The Romans believed that strange creatures called hippocampi, half-horse and half-fish, pulled Neptune's chariot through the seas. One thousand years ago, the Norsemen sang of Sleipnir, an eight-legged horse ridden by Odin. Images of these bizarre beings can be found in Egyptian wall paintings, Greek statuary, and Roman mosaics.

The great philosopher, Plato, has a credible claim as the first articulate genetic determinist, arguing that to ensure the best possible society it was essential that the most talented persons have a disproportionately large number of children. Upper-class Romans were obsessed with family ties and the power of nature over pregnancy. But, as with so much else, during the long cultural quiescence in Europe that followed the fall of Rome (5th to the 8th centuries), there is virtually no evidence of interest in heredity. Perhaps the closest (a reach, I admit) is the persistent human fascination with grotesque bodily forms recorded in the bestiaries, illustrated manuscripts featuring exotic animals. From the 6th century until the Renaissance, the Catholic Church, the sole institution (in the West) that was systematically propagating knowledge, used bestiaries to inculcate moral values. It taught that an animal's physiognomy carried morality tales. By the 13th century, the illustrated bestiary was an important art form; at least five were created in the monasteries of France during that century. During the 13th and 14th centuries, the monks' depictions of animals became more realistic, which may reflect the first stirrings of zoology.

Perhaps the most widely recognized representations of physically abnormal beings are the gargoyles. Although they were occasionally used in Roman times (often as the mouths of waterspouts), the animal shapes that still adorn the great cathedrals of Europe were commonplace in Romanesque and Gothic buildings. In late medieval and early Renaissance times, gargoyles rarely served an architectural purpose; they were used to ward off evil spirits—sort of spiritual scarecrows. Although we think of gargoyles as imaginary beasts, many of the cathedral figures are prosaic lions, dogs, and humans. Although most gargoyles were intended to protect, others appear to admonish about the wages of sin. Hundreds of years ago, humans thought birth defects were the work of the devil. Gargoyles in the shape of deformed humans might have served as a warning about the consequences of immorality.

As art revived during the Renaissance, the artistic focus on biblical themes expanded to include more earthly images. During the second half of the 15th century, Leonardo Da Vinci, one of the world's great anatomists, drew many portraits of men and women with distorted faces and malformed limbs. He also studied human fetuses and the structure of the uterus. From the 17th well into the 19th century, favored painters produced magnificent portraits of the dwarves who were in the court retinue of every capital in Europe. First among them was Velasquez, whose portrait of the court dwarf, Don Antonio el Ingles, hangs in the Prado in Madrid. Artistic fascination was not limited to dwarves; in parks in Florence, Rome, and Siena one can still find old statuary of human figures with the stigmata (flat face, small head, large, protruding tongue) of Down syndrome, and a few with features suggesting other congenital defects.

Throughout the 19th century, the first in which science was ascendant, many writers chose themes involving human monsters. Mary Shelley's *Frankenstein* (1818), written partly in response to the public's fascination with the growing ability to harness electricity, rightly marks the beginning of a literary tradition that led to H.G. Wells's *The Island of Dr. Moreau* (1896), Bram Stoker's immensely popular *Dracula* (1897), and Robert Louis Stevenson's *Dr. Jekyll and Mr. Hyde* (1910). All four novels involve dramatic transmutations of the flesh caused by external forces.

In the middle of the 19th century, P.T. Barnum was among the first cultural impresarios to recognize the public fascination with human oddities (most due to genetic errors). From 1841 to 1865, Barnum's American Museum, which displayed many "freaks," was a favorite New York City destination. Barnum grew rich by using first the museum and then the traveling circus to present bizarre variants of the human condition, most famously, the diminutive (28 inches tall) Tom Thumb, to an enthralled public. Tom Thumb was the stage name Barnum gave to Charles Sherwood Stratton, the son of a carpenter in Bridgeport, Connecticut. He probably had a genetic disorder called primary hypopituitarism (the pituitary gland produces growth hormone). Barnum's marketing genius reached its zenith when he turned the 1863 marriage of Tom Thumb and the equally diminutive Lavinia Warren into a major social event in New York.

During the 1880s, Joseph Carey Merrick, the "elephant man," was a London celebrity. Merrick's head, right arm, and right leg were grotesquely enlarged. He thought his condition was the result of his mother's fright at seeing an elephant when she was pregnant. After his mother died (when he was 12), his stepmother rejected him and he was forced to live on his own. As his disability worsened, he turned to sideshow work with limited success. Merrick then sought and was given a home at the Royal London Hospital. During his four years there, Merrick was befriended by members of the royal family, some of whom regularly visited. Toward the end of his life, Merrick, who may have had a genetic disorder called Proteus syndrome, could only breathe by leaning forward with his head on his knees. He died of asphyxiation in 1896 at age 46.

During the 1930s, persons with genetic conditions that caused extremes in height, weight, head shape, hairiness, and skin color were the star attractions of the sideshows. Poor souls with odd disorders such as Seckel's "bird-headed" dwarfism were immortalized in film by Todd Browning's classic, *Freaks*, in which they revolt against an evil sideshow manager. Several decades later (1978), the literary critic, Leslie Fiedler, wrote perhaps the definitive work on the role of the grotesquely human in American art and culture. Fiedler just missed the chance to record the early impact of molecular genetics on American culture.

DNA and Modern Art

In the last 50 years, artistic and literary exploration of the power of genes has become much more sophisticated. Fascination with sports of nature has given way to the mesmerizing effects of molecules. In 1953, James Watson and Francis Crick deduced the structure of the double helix, which has become an icon of modern science. Crick was the first person in the world to sketch a double helix, a simple doodle-like image that now resides in the archives of the National Library of Medicine. From today's perspective, it is surprising that, despite the immense importance of the discovery, DNA chemistry did not immediately generate a flood of scientific publications. Nor, with a couple of exceptions (notably, two paintings by Salvador Dali in 1957 and 1963), did the now ubiquitous image quickly capture the artistic imagination. This is likely because the immense impact of DNA on our sense of self could only be appreciated after scientists had developed the tools (such as restriction endonucleases and the polymerase chain reaction) to manipulate it. This happened in the 1980s. The very tools that enabled the Human Genome Project also stimulated a field that we could call "genomic art." Today, images of DNA and the advances that it has enabled are common in painting, sculpture, and literature. Far more often than not, artists use the images to express fear (and sometimes, loathing) about the juggernaut of molecular biology.

Several years ago, Suzanne Anker and Dorothy Nelkin argued persuasively that "genomic art" encompasses musings on five themes: the reduction of the body to a molecular program (DNA sequencing), the meaning of mutation (the molecular basis of monstrosities), the blurring of species boundaries (transgenics), the notion of perfection (genetic modification of human embryos), and the commodification of nature (genetically engineered food). The overarching theme is anxiety that molecular genetics has presented men with tools of unimagined power—tools with which they can reshape nature. Sometimes, the artist outruns reality.

In 1994, Thomas Grunfeld painted *Misfit*, a portrait of a St. Bernard dog with a sheep's head (strikingly reminiscent of Egyptian deities!). Just a few years later, scientists managed to create a mouse with cartilage shaped like a human ear growing out of its back! While most people correctly view the current debate over use of frozen human embryos in stem cell research as the latest skirmish in the abortion war, a deeper discontent may be the fear that someday we will have the tools to enhance the genomes of individuals. They worry that George Orwell's thought police, conjured with such frightening effect in the celebrated novel, *1984* (1949), will be joined by genetic police as in the film, *GATTACA* (1998). Such interventions may never become sanctioned by the state, but enabling technologies could emerge.

The current influence of molecular biology on pop culture is immense. The post-Hiroshima, radiation-induced monsters of the 1950s and 1960s have given way to superheroes such as the X-Men and Spiderman, who acquire their powers through mutations in their somatic (cellular) DNA. Godzilla, the quintessential monster of the 1950s, has been superannuated by the dinosaurs in Jurassic Park (1997), which were created by genetic engineering. Sam Raimi, the director of *Spiderman*, knew that in the original 1962 comic book, Peter Parker is bitten by a radioactive spider. With the success of the Human Genome Project, Raimi thought it made more sense to have a spider's poison change Parker's somatic DNA, causing his body to make web glands. Lost in transition is the fact that radiation is the quintessential cause of mutations in DNA.

Incidentally, in 2003, a judge at the U.S. Court of International Trade in New York ruled that the X-Men are "nonhuman creatures." The decision set off a howl of protest in the world of action comics. Chuck Austen, the author of Marvel Comics' *Uncanny X-Men*, was quoted in the *Wall Street Journal* as saying he had spent years emphasizing their humanity, to show "they're just another strand in the evolutionary chain." How did the good judge get thrust into such a ticklish spot? If the plastic X-Men figures (almost all of which are manufactured in Asia) are legally regarded as "creatures," under current law they carry a lower import tariff than if they are categorized as "dolls," which are defined as "human" figures.

Ira Levin's novel, *The Boys From Brazil* (1978), later a film, was the first successful thriller on the theme of human cloning. The cloning genre, now well established, includes many novels, including those by Dean Koontz (*Mr. Murder*) and Robin Cook (*Seizure*), with Kevin Guilfoile's *Cast of Shadows* (2004) and *Never Let Me Go* (2005) by Kazuo Ishiguro being the latest additions. On May 9, 2005, the *New Yorker* commemorated Mother's Day with cover art by the celebrated graphic designer, Anita Kunz. The cover depicts a flaxen-haired young woman dressed in fine silks reminiscent of the Elizabethan era. On her lap sits a young child whose features indicate she is a clone of the woman. In her left hand, which covers the clone's heart, the woman holds a test tube filled with a mysterious liquid. Of course, the adult is not the mother of the young clone. She is an older identical twin!

Genetics has been a major Hollywood theme for some time. In 1992, Ridley Scott directed *Blade Runner* (one of my favorite sci-fi movies, starring Harrison Ford), a dark meditation on Earth's future at a time when genetic engineers create pets to order and the line between genetically engineered robots (creatures called replicants) and humans is so blurred that the protagonist, hired to kill a group of renegade replicants, falls in love with one. Over a span of two decades beginning in the late 1970s, the four *Alien* movies starring Sigourney Weaver cleverly mixed space travel, hostile aliens, and DNA. In the Star Wars films, the Force has a biological basis and is inherited. In the Harry Potter films, wizards usually inherit their powers, and they generally have a longer life span than do "muggles." In the most recent (2004) Pixar film, *The Incredibles*, Mr. Incredible and Elastigirl have three children, each of whom has inherited a different set of superpowers.

On television, the medium with the most insatiable need for new plots, the "genetic gee whiz" genre is also well established and growing steadily. Episodes based on genetic themes such as non-paternity and undisclosed adoption have been standard fare in soap operas for several decades. Ever since the O.J. Simpson trial, crime shows have routinely woven DNA testing into the story lines. Recently, the far-reaching possibilities inherent in the ability to manipulate DNA have emerged. In 2003, CBS launched a TV series called *Century City* that was built around the cases handled by an LA law firm in 2030. One show features a legal battle over cloning humans; another is a malpractice suit for failure to warn that a child would be born gay.

Genetic manipulation, which has long been standard fare in the genre of science fiction, continues to be so. Justina Robson's latest novel, *Natural History* (2004), depicts a future in which the human race has been split in two through genetic engineering. Over the last two decades, immensely successful writers, such as Michael Crichton (*Jurassic Park*), Robin Cook (*Mutation*), and Ken Follett (*The Third Twin*), have sold millions of beach novels centered on the abuse of genetic technology. More subtle writers, such as the English detective master, P.D. James, and the New Mexican novelist, Tony Hillerman, have used DNA forensics to solve crimes. In 2004, Ian McEwan, the fine British writer, published a gripping short story in the *New Yorker* called "The Diagnosis," which turns on the recognition by a surgeon that a mugger is in the early throes of Huntington's disease. In the same issue, the American writer, A. M. Homes, published "The Mistress's Daughter," in which DNA testing to determine paternity is a key element.

In no small part because of the thoroughness with which it has penetrated the best-seller lists, comics, and movies, the public perceives DNA technology as a potential tool for evil as well as a tool for good. In 2002, Caryl Churchill, a highly regarded playwright, wrote *A Number*, a stark, two-actor play about the ramifications of a man's decision to clone his infant son after both wife and child were killed in an accident. *A Number*, in which Sam Shepherd plays the lead, had a solid off-Broadway run in New York City. One wonders whether these productions influence ethical debates that spill into the nation's legislative halls.

DNA AND THE PRESS

The most persuasive evidence of the impact of genetics on our society is its impact on everyday language. One simple way to measure that is to read the newspapers. As Peter Edidin, writing for the *New York Times*, put it: "The ideas and images of the Genetic Age...have burrowed so deeply into human consciousness that it is hard to imagine a world before them." Politicians and pundits, novelists, poets, journalists, essayists, sportswriters, and ad copywriters now regularly use words like DNA and genes to signify innate talent or predisposition to behaviors that would be extremely difficult to modify. I could offer hundreds of examples, but I will restrain myself. Here are just a few of my favorites.

According to the *New York Times*, Senator Kennedy recently warned fellow Democrats not to give in and become "Republican clones." In a human interest article on families that have a passion for collecting, *Times* reporter, Ralph Gardner, Jr., quotes George Wachter, the worldwide head of the old masters department at Sotheby's (whose son is a passionate collector) as saying, "I think it's in people's genes." In a recent interview with Walid Jumblatt, leader of Lebanon's Druse community, reporter Michael Young asked him why he was taking such physical risks. He answered, "We like to be free. Maybe it's genetic." Writing in the Bridge column in the *New York Times*, Alan Truscott waxed eloquent on the successes of the extended family of Gail Greenberg playing championship bridge. Speaking of a new member of the family, Truscott wrote in June of 2004 that "Julian Piafsky, who was born in February in New York City, has exactly the right genes for bridge success." Truscott reported that the infant's parents and grandparents were already discussing "what system he will play." Truscott did not point out that in the case of championship bridge, nurture almost certainly trumps nature. One is not born with an innate ability to play great bridge. However, the odds of becoming a great bridge player are substantially improved if one is taught daily from early childhood by several champions.

The suggestion that someone should be cloned can be offered as the ultimate compliment. Again in the *Times*, a business reporter, celebrating the best feats of 2004, gave highest kudos to Kirk Kerkorian (the financier who specializes in owning Las Vegas casinos) for one of his acquisitions, concluding, "You should write a book, clone yourself and take a vacation." Astute copywriters and marketers have also been quick to appropriate the DNA mystique. A perfume called DNA hit the market several years ago. Jewelers have incorporated the double helix into bracelets. One ad notes that the bracelet designer "represents the beauty of the DNA molecule using intertwining strings of sterling silver beads in these gorgeous pieces." Taking the DNA magic one step further, a company called GeneLink has offered to store a person's DNA indefinitely, as a sort of biological heirloom. No one anticipates the ability to reconstruct a human from DNA, but the idea is a lot less distasteful than the current cryogenic offerings that garnered so much press after the death of baseball great (and my childhood hero), Ted Williams.

Both print and television advertising now routinely use images of genetics. In its magazine ads, the Eastern Mountain Sports Company now uses the term "Y chromosome sizes" and "X chromosome sizes." In 2004, an ad in the *New York Times Style Magazine* began: "In this brave new world, where science has enabled short people to become tall and homely people to become 'pretty,' isolating the style gene is only a matter of time. Until then, all those who wish to become comme il faut must rely on accepted norms—and stylists... ." The ad, I hope, is meant to be humorous, for it lacks any scientific basis. If treated for several years with regular injections of growth hormone, children with any one of several uncommon genetic growth disorders are able to reach a modestly greater height than they would have without treatment. The expensive treatments change them from being extremely short to being quite short. Genetically engineered drugs cannot make anyone tall. I do not want to seem cranky, but the level of scientific misinformation conveyed by such ads is worrisome.

The automobile industry and its analysts also have a special attachment to genes. One of the more clever TV ads in 2004 opens with a tennis game between an athletic-looking adult man and a young child. Strangely, the man is no match for the child. The ad closes with the child running off the court to his dad who has just driven up in a beautiful new luxury car. The dad is Andre Agassi, the tennis star. The ad suggests that the ability to achieve greatness in tennis and the ability to produce great automobiles are genetically encoded. Writing in 2005 in the *New York Times* about the 9-7X, the first sport utility vehicle from Saab, Danny Hakim, comparing it to other SUVs, used the headline, "Building autos with the same DNA."

Sports columnists, always alert for new verbal imagery, frequently use the DNA metaphor. Writing in the *New York Times* just after the Yankees lost game six of the conference series to Boston, Selena Roberts reflected on the problems of the superstar third baseman, Alexander Rodriguez. After commenting on the infamous play in which Rodriguez was called out for interfering with Boston pitcher, Bronson Arroyo, Roberts compared him unfavorably to the home-grown hero, Derek Jeter. Of Jeter, she wrote, "Born into the league as a Yankee, with a certain DNA from his New York experience, Jeter makes the nuance plays that count for the team—not count against it."

As the election returns were flowing across the nation's televisions on the evening of November 2, 2004, Tom Brokaw spoke of politicians' "political DNA." One wonders whether he borrowed the phrase from Donna Brazile (Al Gore's campaign manager in 2000) who, in an Op-Ed column in the *New York Times* on September 19 of that year, wrote, "For some mystical reasons that can be explained only by dissecting his political DNA, Mr. Kerry is like the prizewinning horse Seabiscuit—he runs best from behind." Even President Bush has embraced DNA iconography. In an interview in January of 2005 concerning a book about freedom and tyranny that Bush liked, he told a reporter that its themes resonated with "my Presidential DNA!" Reluctantly, I acknowledge that George W. Bush is the only person who has even a remote right to use the phrase. He is the son of a President.

The influence of DNA on journalism extends even to the usually staid financial press. Writing in the *New York Times* about problems in the mutual fund industry, reporter Diana B. Henriques quotes Michael F. Price, a legendary fund manager, as saying that the industry's culture has been hijacked by "outsiders who lack the genetic code required to manage money for other people." In reviewing *Winning Sure Beats Losing*, a self-help business book, *Times* writer, Ginia Bellafante, after rating herself by the book's metrics, asserts that, "Clearly I have a winner's DNA." In 2003, an article in the *New York Times* on the misfortunes of Seagram heir Edgar Bronfman, Jr. (who had blown a couple of billion dollars of family money) quotes Peter C. Newman, who wrote a book on the fabulously wealthy family: "Never has a gene pool dried up with more drastic consequences."

I suppose it was inevitable that genes would make it into the *New York Times* feature called *Modern Love*. Early in 2005, it ran an essay entitled, "His genes hold gifts. Mine carry risk." The essay is by a young woman who has a one-in-two chance of carrying an X-linked gene that would cause a severe skin disease called ectodermal dysplasia if it were passed on to a son (daughters are not affected because the gene on the other X chromosome neutralizes the problem). She describes her husband-to-be as nearly perfect—having "uncommon athleticism, shining health, beautiful teeth, sharp concentration, perfect aim (he was a college basketball star)." She says, "Dan has genetic gifts he wants to give." Although the woman is a fine writer, she got some key parts of her story wrong. Although ectodermal dysplasia (ED) is a single-gene disorder that can be diagnosed and defined, athleticism and beautiful teeth are due to the interaction of many genes with many environments (I bet Dan wore braces). When she and her husband have children, it is just as possible that they will be like her (with her apparent lack of athleticism and less than beautiful teeth) as it is that they will be like him. Furthermore, it is possible that they both carry some other mutation that will cause a completely unexpected health problem for one of their children. The reality is that every egg and every sperm carries "risks." Fortunately, most of them do not come to fruition. The essay is really about the quandaries of modern pregnancy generated by the ability to learn and react to knowledge about the future of a child. Although the author did not disclose whether she would make use of preimplantation genetic testing to avoid bearing a son with ED, many women would. As her affected brother said, "I wouldn't want your kids to have it."

How powerful is the potential impact of genetic determination on mass culture? A recent book by molecular biologist, Dean Hamer, *The God Gene*, argues that we are genetically programmed to have faith in a supreme deity. The image of the DNA molecule is now used to convey a sense of predestination and inevitability. I doubt that Hamer's argument that people go to church for the same reasons that beavers build dams—because their genes make them—will have much impact on attendance.

How might public fascination with DNA manifest over the next decade? Besides the production of ever more television plots and blockbuster films, one good bet is genetically engineered pets. In a reprise of *Blade Runner*, which presciently portrayed designer pets, in 2002, a Florida company became the first to market a genetically engineered pet—the GloFish—a goldfish which contains a foreign gene that codes for green fluorescent protein, and which makes it glow in the dark. California lawmakers have been asked to prohibit the sale of GloFish on grounds of cruelty to animals! Elsewhere in this book, I report the first commercial creation of a cloned cat. It was $50,000 in 2004, and by March of 2005, it had dropped to $32,000. Still pretty steep, but I bet it will fall by a factor of 10 over the next few years. With a price tag of $3,000, the novelty value might well drive the cat cloning business. It would certainly support a dog cloning business. People routinely pay $1,000 for a thoroughbred puppy.

The double helix is a central icon in our post-modern society. Although it originally symbolized our power to discern the most secret facts of life, with each year it symbolizes ever more forcefully our growing ability to alter life's essence. Humans have been changing the planet for millennia. For at least several hundred years, we have been radically affecting the course of evolution. But everything we have done to date pales before the reality that we will someday be able to reengineer ourselves. Genetic themes in art and genetic metaphor in language reflect our uneasy accommodation to that awesome reality. We can anticipate more dramatic reflections in the future.

Is it ethically permissible to conceive and bear a child so that he can provide bone marrow to save the life of a sibling?

19

Preimplantation Genetic Diagnosis

A few years ago, a couple of weeks before Christmas, I took a break from my writing and went to the post office to mail some presents. The clerk asked me if I wanted to insure the packages, and I declined. While calculating the postage, he told me that December 20 was the busiest mailing day of the year. That got me thinking about the millions of packages moving across the country, and wondering what—ounce for ounce—is the most valuable item sent through the U.S. Mail or a private courier service.

I think I know: a certain kind of human cell. A human cell weighs less than a millionth of a gram, which certainly helps my argument. But, except in the most unusual cases, cells have no value, so how can I justify my claim? I am contemplating a most unusual cell, the genetic analysis of which decides whether an early human embryo becomes a human being or is destroyed. This cell has unique value because the couple that used in vitro fertilization to create the embryo from which it is derived has a one-in-four chance of having a baby with a severe genetic disease. They seek a pregnancy that will not carry that risk. By obtaining genetic analysis on a cell from each of several embryos they create, the couple seeks to identify one that is not affected with the genetic disease.

After gently removing it from the eight-cell embryo, the fertility clinic promptly sends the cell (and those from sibling embryos) to a DNA testing laboratory. The scientists delicately extract the DNA from the cell, amplify it, and test it for mutations. If testing shows that one or more of the embryos are unaffected, they will be transferred to the womb and, hopefully, result in the birth of a healthy child. It costs roughly $20,000 to obtain those few embryonic cells and perform preimplantation genetic diagnosis (PGD) on them. If we assume that as the replacement cost, the little package in which these cells travel has few rivals (by weight) for value.

Preimplantation genetic diagnosis began in 1989 when a British scientist named Alan Handyside reported in *Lancet* that he and his colleagues had combined two new technologies to accomplish an extraordinary first in clinical medicine. They had helped a couple at risk for bearing a child with a serious X-linked disease to give birth to a healthy baby. X-linked disorders are caused by mutations in genes that reside on the X chromosome. Because women have two X chromosomes, they are not affected. A deleterious mutation on one X is almost always neutralized by the presence of a healthy homolog on the allele. However, in each pregnancy a healthy woman who carries a mutation on one of her Xs for a disease like hemophilia or muscular dystrophy has a one-in-four chance of having a child with the disorder. This risk is calculated by multiplying the one-in-two chance of conceiving a boy times the one-in-two chance of having contributed the X chromosome with the mutation to the egg that was fertilized.

To achieve their feat, Handyside and his colleagues combined the technique of in vitro fertilization that was first used in humans in the United Kingdom in the late 1970s with the polymerase chain reaction (PCR), a chemical method of amplifying minuscule amounts of DNA that was developed in California in the mid-1980s. After harvesting eggs from the woman and fertilizing each with her husband's sperm, his scientific team coaxed the conceptuses to divide several times, resulting in eight-cell human embryos. They then performed a microscopic biopsy to remove a single cell from each embryo. In doing so, they correctly guessed that a tiny, eight-cell blastocyst (the technical name for an early embryo) could suffer the loss of one of its cells, yet still develop normally. This was not a wild guess; there was ample scientific basis for it. Next, they extracted the DNA from each cell, amplified it more than a million-fold, and used DNA probes to find out if it had a Y chromosome. Embryos that are destined to be girls would not be affected with the X-linked disorder. By identifying blastocysts destined to become girls and transferring only those into the mother, Handyside could guarantee that the woman would give birth to an unaffected daughter.

Since it was introduced in 1978, in vitro fertilization (IVF) has resulted in about 1 million human births. It has also led to an ever-growing population of frozen human embryos (as many as 400,000 in the United

States), the vast majority of which are stored frozen in IVF clinics with essentially no chance that they will ever be implanted.

Since 1989, preimplantation genetic diagnosis has only been used to assist in decisions leading to about 1,500 births. Why write about such an esoteric technology? There are two reasons. The first is that technical progress suggests that PGD will soon be used much more than it is today. The development of new techniques during the last few years has significantly enhanced the likelihood that very early human embryos can be biopsied successfully and that DNA testing of single cells will yield accurate results. With improved technology, the domain of clinical problems for which PGD is proving useful will expand rapidly. The second is that use of PGD raises important ethical questions. PGD and actions taken based on its results raise issues about the moral and legal status of human embryos, about the prospect of someday using PGD to screen for "desirable" traits (rather than to avoid disease), and about the impact on society of developing technologies that are available only to the tiny fraction of folks who have the resources to use them. Before addressing such topics, a brief overview of the commercial status of PGD may help support my assertion that the time to discuss this new technology is now.

Although he is still a research professor (at the University of Leeds), Alan Handyside is now also the scientific director of The Bridge Center, a major infertility and PGD clinic in the United Kingdom. In the U.S. at the end of 2004, there were at least 26 clinics in 11 states offering PGD, almost always as an adjunct to their IVF services. Since its inception in 1990, the Chicago-based Reproductive Genetics Institute has opened satellite centers in Cyprus, Russia, Ukraine, England, Japan, the Czech Republic, and Belize! Specialists in reproductive medicine are steadily expanding the range of problems for which PGD is useful. The Web site of the West Coast Fertility Centers offers GenEnhance (a registered trademark), a gender selection program. If the woman is willing to undergo the considerable inconvenience of therapy to induce superovulation, a couple who can afford it really can choose the sex of their child! In October 2004, Genzyme Genetics, one of the nation's largest genetic testing companies, began offering some PGD services to fertility centers across the nation. Genzyme specializes in studying the chromosomes. In women with recurrent pregnancy loss, fetal chromosomal abnormalities are often the cause. Genzyme's test, which screens the nine chromosomes that are most often abnormal, increases the likelihood that an embryo with a normal set of chromosomes will be placed in the woman's womb.

Ethicists have been discussing the proper uses of PGD for about a decade, but no medical organization in the United States has issued comprehensive guidelines. The American Medical Association Code of Medical Ethics approves of PGD "to prevent, cure or treat genetic disease," and opposes it to select for "non-disease related characteristics or traits." In the United States, the practice of medicine is regulated at the state level; no state has a law or set of regulations that comprehensively govern PGD. England and Australia have set up agencies that have authority over PGD, but they have not yet resolved some of the more ethically controversial policies. As preimplantation genetic diagnosis is likely to grow rapidly in Europe, Australia, and the United States, the need to resolve issues is pressing.

What are the key issues? The threshold moral question is: Should PGD ever be used at all? In choosing PGD, a couple knowingly creates embryos, some of which will not be transferred to the womb. Consider the typical case of a family at risk for a severe autosomal recessive (single gene) disorder. To maximize the chance of having a pregnancy that results in the birth of a child free from the disease, the physicians will harvest about ten eggs, create as many embryos, test all of them, and transfer two unaffected embryos (if there are two) into the woman. Healthy embryos may be saved, but all affected embryos will be destroyed.

Is it ethically permissible to conceive embryos under such technological constraints? Certainly, the goal sought by the couple—to conceive and bear a healthy child—is good. But what of the means to achieve it? Catholic doctrine teaches that it is immoral to isolate procreation from conjugal love, and thus forbids the use of in vitro fertilization as a treatment of infertility. Although no papal encyclical directly discusses PGD, there is no doubt that the Roman Catholic church would condemn a procedure that destroys human embryos. Orthodox Jews and many conservative Protestant sects would almost certainly take a similar position. Most liberal Protestant groups would condone PGD.

The contrasting views turn on two questions: (1) When does life begin and (2) under what circumstances, if any, is it permissible to end a human life? Theological discourse on the beginning of a human life has an ancient history and, not surprisingly, the great world religions do not hold uniform views. The Catholic teaching is that ensoulment (equivalent to the beginning of humanhood) occurs at the moment of conception, whereas Islam teaches that ensoulment occurs 40 days after conception. Neither of these religions countenances abortion. Many liberal Protestant groups sidestep the question of when human life begins, as they must if their theology accommodates abortion.

Until recently, theologians had no reason to ponder the distinction between fertilization and implantation. However, because extrauterine fertilization creates a conceptus that cannot attain full humanhood unless it is placed in the womb, it is possible to argue that its destruction is not tantamount to abortion. Indeed, to argue that an unimplanted human blastocyst has the moral worth of an infant risks defining a human as nothing more than a diploid set of genes, for a frozen conceptus is little more than this. Such a position flirts with biological determinism.

If one can accept the premise that a morally valued human life begins only at the moment of *implantation*, then theological and ethical objections to in vitro fertilization, PGD, and the use of certain contraceptive agents (the "morning-after pill") vanish. However, the decision to define humanhood as beginning at the moment of implantation does not clear the ethical path for approval of PGD, coupled so tightly as it is with making choices about the value of one conceptus over another. As soon as one has granted the point that there are circumstances in which one embryo is more valuable than another (such as in choosing to begin a pregnancy with an embryo not destined to have a fatal disease rather than with one with such a genetic burden), one has agreed that not all human lives are equal. Does anyone have the right to define the criteria upon which to choose the most valuable among embryos? In our constitutional democracy, it is the woman who provided the eggs and who wants the pregnancy—for only she can consent to the placement of the embryo in her body.

Whatever one's theological position on the question of ensoulment or the morality of abortion, one can sympathize with the urgent desire of an at-risk couple to avoid the birth of a child destined to die slowly and horribly or to live a life burdened with severe disability. But where is the boundary that defines the severity of the burden that makes the act of choosing ethically tolerable? No two genetic diseases are alike. Tay-Sachs disease (TSD) is a horrible degenerative disease of which young children die slowly over several years. Virtually every woman who learns that the fetus she is carrying has TSD chooses to terminate the pregnancy. Thirty years ago, most children with cystic fibrosis died before adolescence; today many affected children will live into their 30s, some longer. PGD is now being used to avoid implanting embryos that would be born with cystic fibrosis. Given the progress in caring for individuals with CF, some would question the morality of using PGD to avoid births of children with this disorder. PGD is now also available to older women who wish to avoid parenting a child with Down syndrome. Yet, many couples are willing—even eager—to adopt children with Down syndrome.

For decades, clinical geneticists and obstetricians have steadfastly refused to use amniocentesis and fetal chromosome analysis to sex a fetus in response to similar requests. Social sexing has been condemned as devaluing fetuses of one sex in favor of another, an inappropriate use of scarce technology, and a step toward eugenics. But what is the moral difference between seeking to implant an embryo unaffected with cystic fibrosis and seeking to implant one of a particular gender? Is using PGD to avoid the birth of a child with CF morally acceptable because of the severity of the disorder, its early age of onset, or its limited prospects for cure? Significant gains in treating children with CF suggest that many of the affected children born today have good prospects for living into their middle years, and there is every reason to hope for further therapeutic advances over the next decade.

The first use of PGD was to sex embryos so that a couple at risk for having a child with a serious sex-linked disorder could avoid that fate. As such disorders only affect boys, the selection of female embryos for transfer ensured the birth of a child without the particular disease. Today, a few couples who can afford it are seeking to use PGD simply because they want to have a baby of a particular sex. Imagine the well-heeled couple with three daughters who, nevertheless, yearn for a son. Although the British and Australian regulatory agencies will not permit this use of PGD, a growing number of clinics in the United States will choose and transfer an embryo of the desired sex. This is a sharp break with established tradition, but is it wrong? Many people clearly view such decisions as within a zone of privacy that should not be regulated by government. Many other people view PGD for selecting the sex of a baby as morally reprehensible. One could even imagine that some state legislatures would enact laws (probably unconstitutional) to forbid this practice.

The European Society for Human Reproduction and Embryology views social sexing as unethical, but it does not explain its reasons. The scarce availability of the service, an argument often made in the early days of a new technology, is no longer a credible objection to limiting its application. Indeed, in the United States, social sexing is emerging as just another expensive option for the wealthy, one that is not likely to be any more objectionable than buying expensive automobiles while Africa starves.

One of the most rapidly growing applications of PGD is to identify and avoid embryos that have chromosomal abnormalities, a condition called aneuploidy. The best-known example of this is Down syndrome, in which affected persons have an extra chromosome 21. Abnormalities involving the number of sex chromosomes (which cause comparatively less severe, but still significant, clinical problems) are fairly common (affecting about 1 in 200 live-born children). Aneuploidy is quite common and is a major cause of spontaneous abortion. It is an important possible problem for older (>35) women, infertile women, and couples in which one partner has a type of unusual chromosome called a balanced translocation. For these reasons, there is a growing interest in routinely testing all IVF embryos for aneuploidy. In the near future, it is likely that screening with a panel of 24 probes (one for each human autosome plus the X and Y) will be offered to every IVF couple. This is a standard that in the United States might well be driven by fear of litigation. Given its prevalence among embryos, the severity of the disabilities associated with most forms of aneuploidy, and the existence of a screening test, it would be difficult to defend a clinician's decision not to urge that it be performed.

Some discern especially troublesome ethical issues when a couple chooses IVF and PGD to bear a child for the primary purpose of using him or her as a tissue donor to attempt to save the life of an existing child affected with a fatal disorder. Such babies have been nicknamed "savior children." Between 1985 and 1995, several of these poignant cases, which involved efforts to save the lives of children with Fanconi anemia (FA), an autosomal recessive disorder characterized by bone marrow failure, leukemia, and high risk for other cancers, became national news. Until advances involving PGD, the only effective treatment was bone marrow transplant, a therapy that was far more likely to provide a cure if the cells were donated by a sibling who had exactly the same HLA gene pattern. Unfortunately, for a couple with an affected child, each new pregnancy carries a one-in-four risk that the fetus, too, will have the disease, and the odds of an HLA match are comparatively low. Once it became possible to use PGD to identify the particular HLA genes carried by an embryo, hope for curing children with FA soared. Instead of proceeding pregnancy by pregnancy while time ran out on a dying child, couples could use IVF to screen many embryos for a potential donor sibling.

In 2000, a clinical research team at the University of Minnesota School of Medicine, working with a team at the Illinois Masonic Medical Center, became the first to successfully treat a child with FA with tissue obtained from a younger sibling conceived by using IVF combined with PGD. The Nash family (the couple has permitted public use of their name) had a daughter who was dying of FA. They used IVF to create several embryos, one of which was both an HLA match to the little girl and free from FA. The resulting pregnancy led to the birth of a healthy boy in August of 2000. Several weeks later the team infused blood collected from the placenta and umbilical cord (which contain bone marrow stem cells) into his older sister. Four years later, her blood and immune systems are normal, and she appears to be cured.

The jubilant publicity surrounding the Nash case engendered hundreds of requests from couples around the world with an ill child (the list of disorders which might be cured includes leukemias, lymphomas, autoimmune diseases, severe anemias, and some biochemical disorders) who might also be saved by this new approach. This led some bioethicists to warn about the dangers of what they call the commodification of human embryos. Recalling that in earlier days some couples had aborted fetuses that were not HLA-matched so that they could quickly try again to conceive a donor sibling, the bioethicists asked whether it was proper to use IVF and PGD to create a child for therapeutic purposes. Some also worried that when they eventually learn the reason for which they were conceived, "savior siblings" will be deeply troubled by the behavior of their parents.

I find the first concern ridiculous and the second unlikely. In this age of small families, I often hear parents say that they are having a second child so that the first will not be an only child—as though an only child somehow has been denied something of fundamental importance in the human experience. Parents who decide to add a child to the family in part for that reason do not love the younger sibling any less. Similarly, parents who bring forth savior siblings do not love them any less. They do not offer them for adoption after their bone marrow has been harvested. I think it far more likely that such children become part of a nuclear family that is bound by extraordinary ties.

In the United States, years of ethical debate over the proper use of a new procedure in medicine are often resolved by decisions taken by insurance companies and HMOs about what services they will rule are medically necessary. In 2004, some third-party payers reached rough consensus that PGD may be medically necessary (read "covered") in the circumstances where both partners are carriers of an allele for a recessive disorder, one partner carries a gene for a dominant disorder, or one partner carries a gene for an X-linked disorder. In addition, the disorders in question must be potentially lethal or disabling and have poor treatment options, there must be a reliable test, and the woman must affirm that she does not wish to undergo prenatal diagnosis and selective abortion. Some third-party payers have decided that they will not pay for PGD to screen out embryos destined for late-onset adult disorders such as colon cancer, but most have not issued guidelines. None have agreed to pay for what is called "social sexing."

In addition to covering couples burdened with single-gene disorders, some insurers that cover infertility services have agreed to cover PGD in three circumstances: inability to become pregnant after three IVF cycles, advanced maternal age (the eggs of women over 35 are more likely to have one or more abnormal chromosomes), and when one partner carries a balanced translocation (a normal complement of chromosomes in which two are joined in a way that can often lead to an abnormality in the chromosome complement of the egg).

Insurers do not lightly agree to cover new services. They carefully scrutinize the reports and opinions of relevant professional organizations. Each year, the European Society of Human Reproduction and Embryology, one of the most influential organizations in the PGD field, collects data from the voluntary report of about 25 European clinics. The most recent available report (2002) brought the number of PGD cases up to 1,561. The data show that 370 cases were to screen for single-gene disorders, 334 were for recurrent pregnancy loss, and 78 were for social sexing to balance children by gender. Two encouraging findings were that embryo biopsy was successful 97% of the time and that diagnostic efforts succeeded about 86% of the time. Successful pregnancies were established in about 20% of embryo transfers, not much different from the monthly rate from sexual intercourse.

The European clinics reported 13 patient requests that they had rejected on ethical grounds. These included requests for an embryo that would become an HLA-matched (savior) sibling, requests when the risk of affected offspring was low, a request to identify a male embryo that would have been infertile, and a request to identify and transfer an embryo destined to have achondroplasia (a dominant disorder that is the most common cause of short-limbed dwarfism). The last request is of particular interest because, in making it, the parents inverted the standard notion of "normalcy." A couple in which one parent has achondroplasia was seeking PGD to avoid the birth of a normal child in favor of one who would also have achondroplasia.

In 2007, the Preimplantation Genetic Diagnosis International Society (PGDIS) published voluntary guidelines for PGD clinics. It noted that PGD has been provided to more than 30,000 patients at risk for bearing children with more than 170 different genetic disorders. The PGDIS recognized an expanded list of indications. In addition to couples at risk for bearing children with single gene disorders, it approves PGD to select for embryos that will be tissue matches with existing siblings who are ill and need a bone marrow transplant, and for adults with unusual chromosomal conditions called balanced translocations (that can become unbalanced as the egg or sperm is formed, causing serious birth defects). More controversial was the decision by the organization to recognize that recurrent pregnancy loss and infertility of unknown cause were proper indications for the technology. A recent paper in the *New England Journal of Medicine* reports no benefit of PGD to such couples, a finding that drew a sharp retort from some PGD practitioners that the authors were just not as practiced in the art as they should have been. The rationale to offer PGD to infertile couples is that many may be burdened with recurrent pregnancy loss at a very early stage due to chromosomal abnormalities in the embryo and that PGD can guarantee that an embryo without major chromosomal defects will be placed for implantation.

The most troublesome ethical issue raised by rapid developments in PGD (but which still lies in the future) is whether there should be limits placed on the use of diagnostic techniques to select for positive (as opposed to avoiding negative) traits in children. For half a century, novelists, futurologists, geneticists, and ethicists have worried that our ability to decode the human genome ordains a future in which we can first pick, and ultimately, shape important traits in offspring. No one has raised this concern more effectively than Aldous Huxley in his prescient novel, *Brave New World* (1932). In the chilling opening chapter, the Director of the Hatchery in London explains Bokanovsky's process, an extrauterine technology that combines cloning with embryo manipulation to create up to 96 identical individuals, each fit for a particular station in socioeconomic life. In Huxley's world, many human embryos have substantial cognitive limits which will make them comfortable in performing sets of repetitive, uninteresting, but essential tasks (please substitute here your own least favorite job) in the society.

I do not fear that we will soon be using a technology reminiscent of Bokanovsky's process to create drones. But it would not surprise me if by mid-century, wealthy couples routinely use PGD to accomplish much more than avoiding the birth of a child with a severe, untreatable disorder. PGD is already being used to identify embryos at risk for late-onset adult diseases such as Huntington's disease, polycystic kidney disease, and colon cancer. If one is taking the trouble to avoid giving birth to a child with one severe disorder, does it not make sense to avail oneself of tests to screen out scores more? Currently, there are not enough available cells to obtain the DNA needed for such mass screening. However, Alan Handyside, the first to successfully perform PGD, has managed to grow hundreds of cells from a single cell plucked from an eight-cell mouse embryo. If that could be done routinely with human cells, it would be possible to develop a much broader approach to screening human embryos. Such a service would be offered to the tens of thousands of couples who each year use IVF to circumvent infertility. The rationale: You have worked so hard to have a child, should we not use tests to rule out unanticipated risks and select the embryo that appears to have the best chance of being the healthiest child?

The use of PGD to screen for gene variations that appear to enhance socially valued traits (intelligence, musical talent, physical prowess) is a long way off, and will surely engender debate, but in a free-market economy, it is disingenuous to argue that such screening would be prohibited. Sure, geneticists will subject early versions of such screening to withering criticism on scientific grounds (they will be partially correct), liberals will denounce the tests as further evidence of the growing socioeconomic stratification, and insurers will refuse to pay for the service. But these are the very arguments that were made against IVF, a procedure that over the last 25 years has led to the birth of 1 million children! Since the most highly prized traits are the products of complex interactions between several or many genes and the environment, it will be for the foreseeable future impossible to offer promises based on test results. Rather, the testing service will act more like a bookie, giving the odds on which of several embryos is more likely to have one or more prized traits. Still, people spend lots of money for far less important goods and services.

It would be naive to say that there are insurmountable technological barriers to expanded PGD screening. Handyside and others are already at work on a project that might be appropriately called the "embryo chip." The concept is to use advanced DNA testing techniques to ask hundreds of genetic questions about an embryo's future health risks. The major challenge is not acquiring DNA data; it is building the clinical database that experts can use to interpret risk. The degree of difficulty will vary widely depending on the question that is asked. For example, it should someday be possible to screen embryos for most of the rare disorders for which we currently use newborn screening. On the other hand, the notion of using embryo chips to alert to risks for common cancers and coronary artery disease, disorders heavily influenced by the environment, is remote.

In predicting the future, people usually underestimate the power and impact of technology. For example, virtually no major science fiction writer in the 1940s or 1950s anticipated the widespread use of a personal computer or the Internet. I will go out on a limb and predict that in four decades, wealthy individuals in technologically advanced societies will have the option of obtaining detailed genetic evaluations of embryos that they create in vitro. Access to such technologies will allow them to cling to the illusion that they are choosing the best possible baby, the baby most likely to fulfill their dreams. Why do I call it an illusion? Because environmental influences and the ever-present role of chance will often shape the future child's life trajectory in a manner that is impossible to predict. Still, eugenics is, I think, a part of the human future. There will be rancorous public debates about ethical questions, but they will not much affect the outcome. Our future includes genetic technology harnessed in the service of reproductive choice.

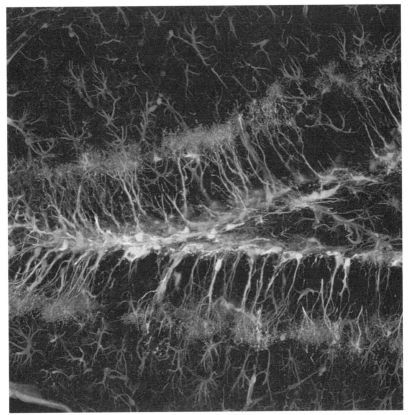

Stem cells in the nervous system. (Reprinted, with permission, from Mignone et al. 2004.)

Stem Cells

It takes about 24 hours for a sperm cell to penetrate the wall of an egg, thus uniting the two sets of genetic information in a new cell called a zygote. Fertilization occurs near the ampulla (opening) of the fallopian tube, the spot where the newly ovulated egg is assaulted by the horde of writhing sperm cells. A newly formed zygote weighs less than 1/100th of a gram (about 1/7,000,000th of an adult human). The zygote's main task is to make the perilous (and often fatal) journey to the uterine wall where implantation occurs. The zygote divides for the first time about a day after fertilization. Viewed through a powerful microscope during its second day, the human zygote looks like the face of an owl—each of the two nuclei has a sharply defined, rounded border, and together they resemble a pair of closely set eyes. Identical twins come into being when this first round of cell division is complete, for each of the two new cells is pluripotent. Each nucleus contains all the genes needed to produce a complete human, and those genes are still programmed to do so.

During the next few days, the zygote divides several times, giving rise to a blastocyst. Through a phase contrast microscope, a human blastocyst looks much like a full moon on a clear night. The unevenly layered cells give the blastocyst a cratered topography. In cross section, the blastocyst looks like a hollow cylinder with a tiny mass of cells concentrated along the rim of one quadrant. All these cells are embryonic stem cells. They are not yet committed to become a particular organ, and each is (if properly isolated) still capable of launching a new human being. It takes nearly a week for the blastocyst to reach and burrow into the uterine wall. Once anchored, it splits into two layers and gives rise to an embryo. Just before it implants, a healthy human blastocyst is a mass of 80–160 cells with a diameter of 4,500 microns (roughly the height of the letter i). The primitive streak—the line of cells that will become the body's axis of symmetry—organizes about 14 days after fertilization. The heart begins to beat on the 22nd day.

Developmental biologists have long known that cells in some organisms have the ability to regenerate body parts. We are all familiar with the ability of a starfish to grow a new limb. Research on how to generate a whole organism from a single somatic cell, the process known as cloning, has been ongoing for many decades. Plant biologists cloned carrots more than 50 years ago. In 1953, Leroy Stevens, a scientist at the Jackson Laboratory in Bar Harbor, Maine, was working on a project funded by the tobacco industry, to investigate whether it might be the paper in cigarettes, rather than the tobacco, that caused cancer. Some of the mice he was studying developed extremely unusual tumors in their testes. When Stevens examined these tumors, he found that they contained teeth, hair, and other well-formed tissues. He named these germ-cell tumors teratomas (from the Greek for monstrosity). In subsequent research, he discovered a "pluripotent embryonic stem cell," a tumor cell that could give rise to different kinds of tissue. But he only could isolate these amazing cells from tumors.

The next big advance came during the 1970s, when J.B. Gurdon and other developmental biologists produced adult frogs from single cells dissected from frog embryos. However, even with this advance, the accepted wisdom was that the genetic program of an adult *mammalian* cell could not become the progenitor of a new animal. Scientists believed that the genomes of such cells were so highly differentiated that they could not be reprogrammed to encode the message of a cell in an early blastocyst.

The cloning of a sheep (Dolly) in 1997 by Ian Wilmot and his colleagues at the Roslin Institute in Scotland revolutionized developmental biology. In essence, Wilmot demonstrated that when a nucleus from a mature cell (say, a skin cell) is transplanted into an enucleated egg, proteins in the cytoplasm of that egg have the capacity to reprogram the genes in the skin cell nucleus to occupy nearly the same on and off positions that its ancestral cells had when they were part of an early blastocyst.

In the few years since Dolly burst upon the scene, scientists have successfully cloned many species (including horse, pig, rat, mouse, cat, cow, and dog), but they have also found that cloned animals often have serious abnormalities. Leading scientists, such as Rudolph Jaenisch, one of the major figures in the development of gene transfer ("transgenics") at the Whitehead Institute in Cambridge, Massachusetts, doubt that cloning will prove to be an acceptable tool in animal husbandry or a clinical option in human reproductive medicine. In his view, even a few subtle reprogramming errors among the thousands of genes in the genome render the possibility that a cloned individual will lead a healthy, vigorous life remote. Because of the strong likelihood of physical harm, he considers the idea of reproductive cloning in humans to be morally reprehensible.

In November of 1998, James A. Thomson, a cell biologist at the University of Wisconsin, became the first person to isolate "stem" cells from human embryonic tissue. His paper, which elicited vast press coverage, was an important extension of research that Martin Evans, a professor at the University of Cardiff in Wales, and his colleague, Dr. Matthew Kaufman, had been pursuing for almost two decades with mouse embryos. In the early 1980s, Evans demonstrated that one could tease apart the undifferentiated cells in an early mouse embryo and grow them into particular types of cells, such as heart muscle. Evans was awarded the Mary Lasker Prize, often called the American Nobel, for this research.

The term, stem cell, was probably first used consistently by Gail Martin, a scientist in San Francisco, who in the 1980s was studying the development of mouse embryos. Today, scientists agree that a stem cell is one that (1) is capable of unlimited self-renewal, (2) is capable of dividing asymmetrically; that is, to yield one cell exactly like itself and another that is committed to becoming any one of many different cells found among the three cell lines from which mammal organization unfolds (endoderm, mesoderm, and ectoderm), and (3) originates from an embryonic or adult stem cell source. An astounding corollary of this definition is that stem cells in one part of the body, say the blood-forming cells, can, under the right circumstances, become the progenitors of one or more distinct lines, such as muscle or liver cells.

The research of Evans, Martin, and Thomson, as well as that of Dr. John Gearhart of Johns Hopkins University, who (also in 1998) was the first to grow human embryonic germ cells, fundamentally altered our understanding of how organisms are formed and how tissues repair themselves. The discoveries pulled many scientists into the field and engendered countless experiments. Over the last few years, stem cell biologists have shown that embryonic stem cells can be manipulated to become tiny biological factories to fabricate millions of almost any kind of cell (heart, liver, pancreas, nerve) of interest. This has in turn led physicians, especially when contemplating the way in which organ systems fail in old age, to dream of a dramatic new approach to therapy, called regenerative medicine. The therapy of the future, many scientists now believe, will be based not on drugs, but on cells. For example, in 2002, Dr. William Haseltine, then CEO of Human Genome Sciences, predicted a world in which everyone would have access to reserve cell lines to permit them to perpetually rejuvenate themselves. In his words, "The fundamental property of DNA is its immortality. The problem is to connect that immortality with human immortality and, for the first time, we see how that may be possible."

Dr. Haseltine and many others view cell therapy as providing a way to overcome the chronic shortage of spare parts imposed by the immune system barrier. They foresee a day when individuals will bank their stem cells. Frozen in tiny vials, these cells could be thawed, expanded, manipulated, and delivered to the body to satisfy any need it might have. For example, stem cells could be coaxed to become islet cells for diabetics and the appropriate brain cells for persons with Parkinson's disease. If returned to our aging or diseased tissue at the right time, these cells might reset the clock for the tissue or organ that was failing. Since 1998, a large research effort has generated much hope that in the not-too-distant future, physicians will use cell therapy to rejuvenate heart muscle, cure juvenile diabetes and Parkinson's disease, and ameliorate Alzheimer's disease.

One important area of research in this new field has focused on detecting stem cells in tissues where, heretofore, they were not known to exist. In 2000, several teams discovered that our largest organ—the skin—steadily produces new skin from stem cells that work as tiny cell factories inside hair follicles. The hunt is on for many other types of stem cells. Another important item on the research agenda is to learn more efficient methods to direct stem cells to become particular types of differentiated cells. In 2001, Dr. Ronald McKay, a scientist at the National Institutes of Health (NIH), figured out how to make mouse embryonic stem cells differentiate into a special kind of brain cell that he then used to treat a disease in mice that resembles Parkinson's disease. Dr. Piero Anversa of New York Medical College and Dr. Donald Orlic of the NIH showed that stem cells harvested from mouse bone marrow can be treated with growth factors to convince them to develop into cardiac myocytes (muscle cells). It is possible that, some day, infusions of genetically compatible heart muscle cells could be the treatment of choice to repair muscle damaged by heart attacks.

In 2002, Jaenisch and George Daley, then also at the Whitehead Institute, used a technique called therapeutic cloning to cure a genetic disease in mice. A single-gene defect in the mouse caused a faulty immune system (the mouse version of severe combined immune deficiency). The scientists isolated a nucleus from one of the mouse's cells, transplanted it into an enucleated mouse egg, grew it to a blastocyst, removed some of the stem cells from the blastocyst, transferred a healthy version of the disease gene into the stem cells, treated them with growth factors that caused them to differentiate into mature blood stem cells, and then injected them into the mouse with the genetic disease. The new stem cells quickly expanded, essentially creating a normal immune system capable of providing a normal immune response!

Early on, scientists surmised that in adult mammals there must be as-yet-undiscovered stem cells that throughout life play a key role in remodeling the gastrointestinal tract and the liver, the blood, and skin, as well as other tissues. Recently, it has been established that in adults even the heart and parts of the brain are remodeled under the guidance of stem cells. Those who on religious or moral grounds oppose experimentation with human embryonic tissue (most often obtained from frozen human embryos) have used such discoveries to argue that we should halt embryonic stem cell research and concentrate on learning how to harvest and reprogram adult stem cells. James Sherley, a biologist at MIT, is one scientist who supports such a position. In his view, the use of a human embryo in research is the destruction of one human to benefit another.

In 2004, a team at Seoul National University in South Korea reported a stunning advance in the ability to create pluripotent cell lines from cells taken from human embryos. The scientists obtained 242 eggs that they said were donated by 16 volunteers. After removing the nucleus from each egg and inserting a nucleus from a different tissue taken from the donor, they cultured the cells, eventually coaxing 30 to grow into blastocysts. By dissecting cells from these balls of tissue, they were able to create an embryonic stem cell culture. In 2005, they reported that they had greatly increased the efficiency of their work, and that they had established 11 cell lines, nine of which had been created with tissue taken from people with particular disorders. In culture, the cells appeared to behave as though they could be forced to differentiate into particular kinds of cells, such as muscle, heart, and blood.

Sadly, soon after his second report of a major research breakthrough, Dr. Hwang came under suspicion. In November of 2005, Dr. Gerald Schatten, one of his American collaborators, having discovered that Hwang had violated ethical guidelines by purchasing human eggs rather than obtaining them by donation, abruptly ended their relationship. Hwang admitted the violation, but claimed he had been ignorant of the rules. A few days later, investigative journalists began to raise questions about the validity of the scientific discoveries he had reported. On December 29, 2005, a university committee, that had been convened to assess the situation, reported that the work was fraudulent. Both his famous papers were withdrawn by *Science* (an act tantamount to saying they should never have been published). A few weeks later, Hwang was forced out of the university. In the space of a year, he had risen from relative obscurity to national hero and then plunged into disgrace. His fraudulent actions were deeply harmful to the field of stem cell research, so entangled was it already in ethical controversy.

Fortunately, in 2007, there were two major advances that allowed the field to put Hwang's misadventures behind it. Late in 2007, Jim Thompson, the professor at the University of Wisconsin who had been the first to develop human embryonic stem cell lines, reported a major improvement in his laboratory's ability to establish such cell cultures.

In November of 2007, a team led by Shinya Tamanaka at Kyoto University in Japan reported that it had succeeded in reprogramming adult human skin cells to behave as pluripotent stem cells—that is, progenitor cells that have the potential to differentiate into virtually any of the more than 200 types of cells in your body. Astoundingly, the researchers needed to insert just four regulatory genes (known as *Oct3/4*, *Sox 2*, *Klf2*, and *c-Myc*) into the skin cells to accomplish this feat. The work with human cells followed on earlier research in which they had used the same four genes to accomplish the same feat with mouse cells.

A few weeks later, scientists in Rudolph Jaenisch's laboratory at the Whitehead Institute in Cambridge used Tamanaka's discovery to treat mice that had been genetically engineered to have sickle cell anemia. The scientist snipped cells from the tails, grew them in culture, and used a retrovirus to transport the four regulatory genes into them. After the cells had been reprogrammed to be pluripotent, they used other techniques to command them to become bone marrow (blood precursor) cells. Next, they replaced the defective sickle cell (beta-hemoglobin) gene with its normal counterpart. They then injected these cells into the mice with the disease. The cell line rapidly expanded, creating billions of normal blood cells. Several weeks later, standard laboratory tests showed that the mice had normal red blood cells and normally functioning kidneys. If you think this sounds like the work of wizards, I cannot disagree. Unfortunately, this gene therapy is not yet ready for human trials. The retroviruses that are used to carry the reprogramming genes can disrupt other genes in the cells they enter, and have been shown to cause cancer. Still, it is likely that safe vectors will be developed.

The ability to change adult skin cells into pluripotent cells seems to offer a solution to the strong objections held by about half of Americans that stem cells derived from human embryos should not be used in research. Dr. Jaenisch was among the first to caution that we need to learn much more about embryonic stem cells before we can rely only on induced adult cells as a research platform.

Although every cell in the inner cell mass of a human blastocyst is a stem cell, only about 1 in 1 million cells in the human body is a stem cell. We do not know where most of them reside, and we lack the tools to harvest and study them. Harvey Lodish (another Whitehead Institute scientist), who has devoted much of the last few years to studying stem cells, notes that even with the most sophisticated tools (such as fluorescence-activated cell sorting), his team is often only able to obtain a very small number of blood-forming stem cells from a mouse—far too few to provide material for critically needed experiments.

One of the human diseases for which stem cell research holds out the greatest hope is juvenile (Type 1) diabetes. This autoimmune disorder usually strikes in childhood when the body's immune system attacks the cells in the pancreas that make insulin, killing off most of them before there is any overt sign of illness. Children with juvenile diabetes need exceptionally close medical monitoring and, even with the best of care, as the years pass, affected adults are at risk for blindness and other end organ failures. Shortly after learning about Thomson's success, Douglas Melton, a developmental biologist at Harvard who has spent most of his research career studying frogs, jumped into the field. The impetus was powerful; Dr. Melton has family members with juvenile diabetes. His quest, one strongly supported by the National Juvenile Diabetes Foundation, is to learn how to direct human embryonic stem cells to mature into pancreatic islet cells—the ones that make insulin. Unlike some researchers who are slow to share their re-sources with colleagues, Melton's lab has rushed to establish new human embryonic cell lines and to share them. So far, he has not been able to re-program embryonic stem cells to make insulin, but he is determined.

If stem cells could be commanded in cell culture to become islet cells, it would be a gigantic improvement over the current best treatment for ju-venile diabetes—islet cell transplant. Over the last few years, there have been several hundred such operations worldwide, a minuscule number compared to the thousands of patients. Islet cell transplants require a ge-netically compatible donor, but relatively few families have a relative who matches the patient closely enough to offer a good chance for success.

Ethics and Public Policy

The United States is polarized over the morality of using early human em-bryos in research and cells derived from them to treat patients. Some states have virtually forbidden stem cell research, while in others stem cell re-searchers are creating cell lines from frozen human embryos (more prop-erly, blastocysts) donated by couples that created them via in vitro fertil-ization (IVF). At least one state, New Jersey, has openly bragged (in a full-page ad in the *New York Times*), about its decision to create "the first state-supported stem cell institute."

The reason there are plenty of frozen embryos available for research is that IVF physicians routinely create many for each infertile couple. This maximizes the chances of a pregnancy and spares the woman further surgery to harvest eggs. It is standard procedure to create five or more zy-gotes in the lab, transfer three (in most cases some do not implant) into the woman, and save the rest. Most couples who have children through IVF have extra embryos, and surveys have confirmed that many prefer to donate them to research rather than destroy them.

A central moral question in the debate over embryonic stem cell re-search is, Does a human embryo created in vitro have the same moral sta-tus as a fetus in utero? This is tantamount to asking, When does human life begin? —surely one of the oldest and most contemplated questions in hu-man history.

Over the last two millennia, the moral status of the human embryo has been shaped by a complex pattern of cultural and technological forces.

Aristotle argued that a male human fetus had no moral claim to protection for 40 days (for females, it was 90 days). All four major monotheistic religions have experienced much internal debate on the moral status of an embryo. The Catholic church did not condemn abortion as a crime against nature until the 13th century. Today, it teaches that human life is ensouled at the moment of fertilization. Most Christian fundamentalist groups and the Mormon church (which permits abortion in cases involving rape, incest, or when the woman's health is endangered) agree. Shiite Muslims, conservative Protestants, and many Bhuddists and Hindus also hold this view. Although mainstream Protestant denominations acknowledge that human life is established at fertilization, most argue that personhood is acquired gradually during fetal development. Under Jewish law, an early embryo does not acquire moral status until after it is implanted in the woman's body. Many Jewish scholars consider a fetus to be a potential human life, a being with a potentiality that can be sacrificed to save the life of, or avert serious illness in, the woman. Sunni Muslims believe that an unimplanted embryo is no different from a sperm or egg, a view that accommodates stem cell research. No religion can claim that it has an absolute and unchangeable position on the question of when a human life begins or on the morality of abortion. Even the teachings of the Catholic church have shifted; before the mid-19th century, it recognized circumstances under which abortion was not a mortal sin.

Of all technologically advanced nations, the United States has experienced the most contentious debate over the moral status of human embryonic stem cells. On learning of Thomson's research, Christian conservatives promptly declared that research which involved the destruction of frozen human embryos was immoral. In their view, since the research ended human life, it was tantamount to abortion. Since 1999, a coalition of Catholic and conservative Christian groups has lobbied vigorously to forbid federal funding of the research that Thomson and Gearhart made possible. On August 9, 2001, after months of controversy that included a sharply divided report from an advisory task force he had appointed, President Bush took the unusual step of making a prime-time television appearance to announce his decision. He told the nation that he had decided to forbid the use of federal funds to create new cell lines derived from frozen human embryos, but that he would permit the use of federal funds to support research on existing cell lines.

The solution satisfied nobody. Conservative Christians condemned it as immoral. Leading scientists asserted that existing cell lines were woefully inadequate in number and quality and much too difficult to access. They were right. At the time of the President's decision, NIH claimed to have identified 64 human embryonic stem cell lines in 10 laboratories around the world. However, only 20 of those were in the United States (of which 13 were held by biotech companies and not easily accessed). Of the remaining 7, 5 were controlled by the University of Wisconsin Research Foundation, which had been awarded a patent on them, and 2 resided at the University of California, where they had been created in part through funding from a biotech company. Even more important, it was well known at the time of the President's speech that because many of the existing cell lines were grown in flasks with mouse cells acting as "feeder cells," the human cells would be unsafe for research involving human subjects. In early 2004, only 15 of the 64 originally designated cell lines were potentially available as a resource material to federally funded scientists. In March, Douglas Melton and his colleagues at Harvard Medical School reported that (starting out with 286 donated frozen human embryos) they had been able to establish 17 individual human embryonic cell lines, all capable of differentiating into particular types of tissues. They announced that they would make the cells available to qualified researchers.

The decision by President Bush to sharply limit access by university-based stem cell researchers to NIH funds provoked a storm of controversy. The robust anti-abortion lobby seemed to meet its match as one disease advocacy group after another testified before congressional committees, demanding federal support of research with human embryonic stem cells. The Juvenile Diabetes Research Foundation was especially forceful, planning at one point to have every child with diabetes send a scrapbook about the illness to his or her Congressman.

From 2001 to 2004, the Bush administration was unwilling to loosen the constraints it had placed on human embryonic stem cell research. In response, various states considered, and a few enacted, laws that specifically supported research with cells derived from frozen embryos. In 2003, the California state legislature became the first to provide funds for research involving frozen embryos. In May 2004, New Jersey established a stem cell institute to be run jointly by Rutgers University and the state's medical school. Stanford and Harvard are among the major research universities to declare that they will internally fund embryonic stem cell research. Many others are lining up to follow their lead.

In November 2004, Californians approved Proposition 71. It authorizes the creation of the California Institute of Regenerative Medicine. This public agency will act as an umbrella organization to oversee the expenditure of $3 billion of state funds over 10 years (generated by the sale of bonds) to support stem cell research. Beginning in 2005, California will spend about $300 million a year on this research, more by far than any other country in the world (including the rest of the United States). Proposition 71 will likely make California the world center for stem cell research. Although the conservative Christians who led the fight against stem cell research triumphed in the White House, they will not be able to block all the individual state initiatives. The funding of Proposition 71 is a special threat to stem cell research in Massachusetts. In 2005, Governor Romney announced he would introduce a bill to limit ES research to material derived from donated frozen embryos, and to prohibit the deliberate creation of human embryos in the test tube. If the liberal state legislature rejects in it in favor of a broader enabling law, which he then vetoes, it could drive some researchers west.

The debate over the morality of using stem cells derived from human embryos in research has redrawn a number of political boundaries. Most famous is the decision taken by Nancy Reagan, first behind the scenes and then openly, to support such research because of its promise to help in the battle against Alzheimer's disease. Orrin Hatch, the conservative senator from Utah who is a leading opponent of abortion, supports the research. The *New York Times* quoted him as saying, "I just cannot equate a child living in a womb, with moving toes and fingers, and a beating heart, with an embryo in a freezer." Senator Hatch sponsored "The Human Cloning Ban and Stem Cell Research Protection Act of 2003" (S.303) which proposed to outlaw reproductive cloning but to permit some stem cell research. It did not come to a vote. Many in Hatch's conservative home state of Utah were appalled by his stance (a finding that reflects the fact that the Mormon church has opposed work with stem cells), but a poll showed that a majority supported him. Given the growing number of states that are openly inviting stem cell researchers to set up shop, President Bush will not loosen current federal rules that limit use of human embryonic stem cells (HES). Indeed, the federal rules no longer really figure in the controversy.

I wonder how those who believe that an unimplanted human embryo suspended in a tiny vial in a low-temperature refrigerator is morally equivalent to a child suffering from severe diabetes reach that position. Imagine yourself in a building that is on fire and soon will be engulfed in flames. In the room to your left is a wheelchair-bound woman. In the room to your right is a refrigerator that holds a tray of 100 frozen human embryos that you can easily carry to safety. You have only enough time to save the woman or the frozen embryos. What should you do? None of those to whom I have posed this dilemma, including many with strong fundamentalist views and deeply held pro-life stances, have said that they would save the frozen embryos.

Why is the decision so straightforward? On the psychological or emotional level, humans identify much more closely with a fellow person than they do with a potential person or persons. The person at risk in the fire looks like you and has had a set of life experiences roughly like yours. The unformed cell masses in the vials have not. On the philosophical level, the disabled woman is an actual entity, unquestionably a member of the human family, almost certainly capable of crying out for help, and undoubtedly capable of experiencing the agony of death by fire. She is fully human. The embryos in the vial are not sentient. They have no awareness, no capacity for pain, no connection to a family that will grieve over them. They are not persons.

Successful ethical discourse requires a fundamental agreement about the nature of the facts that define the issue. Is it possible that close attention to the biology of human development could support a moral compromise that permits the use of freely donated frozen human embryos as a source of embryonic stem cells? I would like to suggest one possible compromise.

An embryo (whether created through normal procreation or in vitro in a clinic) does not acquire the capacity for humanhood until it implants in a woman's uterus. This has long been recognized in the Judeo-Christian tradition. In both the Old Testament and the New Testament, virtually every reference to new human life makes reference to the womb. Consider these words from Ecclesiastes (11:5): "As you do not know the path of the wind, or how the body is formed in a mother's womb... ." Isaiah (44:24) reads, "This is what the Lord says—your Redeemer—who formed you in the womb... ." Job (31:1) reflects, "Did not he who made me in the womb make them?" Psalm 139 sings, "For you created my inmost being; you knit me together in my mother's womb." The narrative in Luke (2:21) says, "...his name was called JESUS, which was so named of the angel before he was conceived in the womb." A literal reading of scripture supports the view that God's message is that humanhood begins with implantation, the biological fusion of the embryo with the mother.

By the time an embryo arrives at and implants in the womb, it has existed as an undifferentiated cell mass for several days. Unlike Catholic doctrine, which is based on the infallibility of a papal encyclical that holds that human life begins with conception, evangelical Christians do not look to a central authority—other than scripture—to derive their moral positions. If evangelical Christians agree that God's words define humanhood from the womb, then perhaps they could accept the decision by infertile couples, who created embryos in vitro in the hope of having children, to donate those that they will not use for research intended to relieve human suffering.

Assisted reproductive technology ethically requires the creation of more embryos than can safely be placed in a woman's body. To do otherwise would expose the woman to unnecessary pain, risk of harm or death, and cost. Given the uncertain success rates with assisted reproduction, for the foreseeable future there will be a large number of frozen human embryos destined for destruction that could be used to conduct important biomedical research.

There are almost certainly enough frozen human embryos to provide the source of the cell lines needed for most research. If some who are currently opposed to such research on moral grounds accepted "implantation" as the moment of ensoulment, they should be able to accept the use of frozen embryos in research. They should also be willing to accept research involving therapeutic cloning. Cell lines established via therapeutic cloning cannot under any circumstances become human beings. Therapeutic cloning is the research tool that will allow scientists to study cell lines formed with tissue taken from persons with specific diseases. Donated frozen embryos cannot answer that need.

I believe that it is possible to feel respect for a human embryo without according it the full moral status of a child or adult. Humans feel reverence for many living things—the loons on a lake in New England, the giant redwood trees of California, the shrinking remnants of our great prairies—

that are not human. We also generally respect certain inanimate things such as the skeletal remains of persons long dead. The belief that it is moral to use frozen human embryos that would otherwise be destroyed by the couple that created them in ethical research is only impermissible if one believes that the frozen embryo is morally equal to persons reading these words. If one does believe in such moral equivalence, then should not one also oppose the use of in vitro fertilization to assist infertile couples in achieving pregnancy? All such therapies involve the creation and destruction of human embryos. Indeed, must not someone who hews absolutely to the principle that humanhood begins at conception protest current federal policy and insist that all research with human embryos be considered a crime? Are those who oppose research with discarded human embryos really prepared to deny infertile couples the ability to attempt to become biological parents?

Once an infertile couple has successfully completed childbearing, all remaining frozen embryos, the insurance against early treatment failures, are doomed. The embryos may remain in a low-temperature freezer for a time, but sooner or later they will be destroyed. Often when parents lose a son or daughter to head injury, they are asked to permit the removal of the heart, lungs, and kidneys for transplantation. Many agree to make this ultimate gift, no doubt believing that they are giving a special meaning to the life their child lived by adding this act of altruism to his or her personal history. Can we not regard a frozen human embryo in much the same way? Even though it has no chance of personhood, can we not enhance its moral value? Would it not be preferable to be destroyed as part of a research process that might take humankind a step closer to curing a disease than simply to be destroyed?

Nearly 30 years ago, a writer named David Rorvik wrote *In His Image*, a book that purported to be the true story of the first cloning of a human being. The publisher, J.B. Lippincott, inserted a note at the book's front stating that it had no way of knowing if the claim was true. It was a hoax. At the time, no one had published a paper demonstrating that he could manipulate mammalian cells in a manner that would permit cloning.

In 2008, it now appears possible that we are very close to having the technical capability to clone humans. This capacity will flow from the huge scientific enterprise that is being developed to undertake therapeutic cloning. The "Holy Grail" is to be able to remove a few skin cells from a person's body and reprogram them to create a bank of cells that could be used later to treat a disease. For example, scientists could reprogram cells to become insulin-secreting cells for persons with diabetes or to become special brain cells for people with Parkinson's disease. Because the cell lines are derived from the patient, the body would not reject them.

In January of 2008, Stemagen, a small biotechnology company in California, reported that it had created human embryos from skin cells taken from two adult men. Scientists at the company claim that the tiny ball of cells, known as a blastocyst, looks normal. At this writing, they are attempting to derive personalized cell lines from them.

This advance and the many others that will come in the next few years are sure to resurrect the vitriolic debate about the ethics of reproductive human cloning that peaked around 2000, a time when a number of states enacted laws to forbid it. The vast majority of public institutions and individuals are opposed to reproductive cloning. They argue that since it is impossible to know and counter the risks to the baby that would be created by cloning, it is clinically too dangerous to use as a form of artificial reproduction. Further, since the human beings that would be created by research into reproductive cloning would be destroyed in the process, it would be unethical to conduct such experiments. Even if it could be shown that reproductive cloning was no more dangerous to a fetus than is natural reproduction, most people would still be opposed. They sense that a child deserves to be born with his own genetic identity, and that it is wrong for a self-important adult to impose one upon him.

Whether one finds reproductive cloning morally repugnant or not, it is inevitable that someone will create a human being by reproductive cloning within the next decade. Although it is sheer hubris, the desire to be the first to accomplish this amazing feat will ensure that it happens. The student newspapers at the Ivy League Schools routinely carry ads offering up to $100,000 to young women willing to donate their eggs to infertile couples. Couples routinely pay women $10,000 plus expenses to carry a pregnancy for them. It would not be particularly difficult for a scientist to procure the eggs that are needed for reproductive cloning or for the cloning party to hire a woman to carry (probably without her knowledge) the cloned baby to term. I will hazard the guess that the first human clone will be created outside the United States by a private laboratory hired by a wealthy older man who never had children and who has an immense ego.

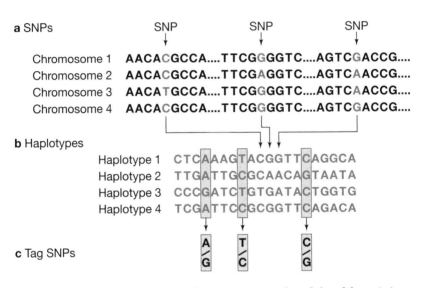

Although humans are 99.9% alike in their DNA sequence, knowledge of the variations (called SNPs) that make up the remaining 0.1% is crucial to the future of medicine. The HapMap Project collected most of these variations by studying populations from Europe, Africa, and Asia. Scientists were able to determine which groups of SNPs were inherited as a unit—a set of markers that represents a long block of DNA. This makes it much easier to study the relationship between variations and risk for disease. (Image courtesy of the International HapMap Project at http://www.hapmap.org.)

Personal Genomics

One of the most important advances in genetics in the last decade was the development of tools (often called "platforms") with which to ask an immense number of questions about a genome (the term we use to describe the complete sequence of a species or an individual) at an astoundingly low cost. These tools do not read the complete DNA sequence. Instead, they ask which of the four letters (A, T, G, and C) that make up the DNA chemical alphabet occupies each of several hundred thousand locations distributed across the more than 3,000,000,000 that comprise a complete sequence. These locations are of interest because research has shown that they vary among the human population. They are called single-nucleotide polymorphisms or SNPs, and the degree to which two people differ at each of these locations is the quintessence of human variation.

To explain the rush to study these SNPs, it helps to recall just a bit of the recent history of genetics. Despite the epic nature of its achievement, the Human Genome Project—which, spending nearly $3,000,000,000 over nearly a decade, completed the first "consensus" sequence of a human in 2001—was able to tell us almost nothing about human variation where it really counts: inside the gene.

Because SNPs introduce variations in the genetic code, some of them alter the nature and function of the protein for which a particular gene provides a biochemical blueprint. Most do not. This is because the vast majority of variation occurs in what is known as the noncoding region of the genome. One of the most mysterious facts about the genomes of almost all higher organisms is that most of the DNA content (~95%) does not code for the production of proteins. Suffice it to say that our genomes are stuffed with various classes of highly repetitive DNA (much of it probably of viral origin) that does not have a clearly apparent function (but, almost certainly, has a function that we have yet to decode).

We are just beginning to understand the role of SNPs in human health and illness. For purposes of this chapter, we should think of SNPs in two ways. Some actually affect the production and function of the proteins that operate our bodies. They do this by altering the level of production (especially SNPs that are in the promoter region of the gene) or by slightly tweaking a protein's shape. Often, such small changes have no effect on the efficiency with which a protein operates, but some SNPs can alter the protein enough to introduce a change that ultimately affects our health. Most do not change the protein; however, because they may be near others that do, they can serve as markers to guide a search, much as the flag in the center of the green 400 yards away is offered to help my pitiful search for a par.

The diversity of SNPs in the human genome is an inevitable by-product of human evolution. Many forces, ranging from cosmic rays to biochemical errors in how DNA recombines during the formation of sex cells, cause small changes in the molecule. If they are deleterious, natural selection quickly purges them; if they confer benefit, they have a chance of becoming part of the species' genome. However, from an evolutionary perspective, most are neutral. They have a fair chance of staying in the genome because they are not subject to the withering forces of natural selection. About 1 in every 1000 of the 3,000,000,000 DNA letters in a sperm or egg is an SNP. These variants are broadly distributed throughout the genome. Of course, no two people have the same set of SNPs, and a child inherits a different set from each parent.

The human genome harbors about 10 million SNPs; because any one of four DNA letters can be at any one location, there can be more than one SNP per location. Fortunately, we will not have to study each one to understand the role of SNPs in human biology. This is because even though DNA molecules recombine as they form the genomes of sperm and eggs, pretty big chunks remain unchanged (i.e., these stretches are not shuffled by recombination). This means that if an unchanged chunk of germ-line DNA contains 50 SNPs, just 1 SNP can be used as a marker for the whole group. If the marker ("tag") SNP is detected, there is a very good chance that the individual who was tested has the other 49 as well.

Let me offer an analogy about the power of surrogate SNPs (several of which taken together are called a haplotype). If you walk through the heart of a big city for the first time, there will be many markers that you can use to infer information. For example, when you walk by a movie theater, you will not have to go inside to know that it has a ticket booth, a popcorn machine, many seats, a big screen, rest rooms, and water fountains. Of course, each of these items will be slightly different in different theaters, but you know from seeing the marquis that they will be inside. Think of the marquis and all the items mentioned above as SNPs. The marquis represents all the items inside that you do not see, but know are there. Obviously, the same analogy could be made with restaurants, banks, department stores, and flower shops. If you think of each as a block of DNA, it will be straightforward to decide what object at each location would properly represent that block. For a bank, it might be the ATM machine next to the door; for the restaurant, it might be the menu in the window.

Early on in the work of the Human Genome Project, geneticists realized that the construction of haplotypes to represent blocks of the human genome would provide a highly efficient tool to investigate the relationship between gene variants and health and disease. They also realized that because of evolution, marker SNPs would differ among populations. Modern humans originated in Africa about 150,000 years ago, and about 50,000 years ago, small groups began to radiate across the planet, creating subpopulations. Modern people of African descent comprise a subpopulation that has had more time in which to experience the evolution of haplotypes. Among Africans, the small blocks of DNA that can be defined by a set of markers are comparatively shorter than those of other groups. As the human family has populated the four corners of the world, each group has become subject to slightly different sets of evolutionary forces and, consequently, has somewhat different sets of haplotype markers.

Even though it was not yet fully developed, scientists realized that they could use DNA microarray technology to discern most of the SNP variations among most of the human population. In 1998, financed by funds from the National Institutes of Health and from comparable resources in several other nations, an international consortium of geneticists organized the International HapMap Project. After much analysis, they concluded that they needed to collect (with consent) DNA from just 270 members of the human family. The Yoruba people of Nigeria provided 30 sets of samples from two parents and a child (called trios). In Japan, 45 unrelated people donated DNA, as did 45 people from China. Persons of northern European orgin provided an additional 30 trios. Using this representative sample of humanity, the HapMap scientists identified common haplotypes. They then delved deeper to identify single spots (called "tag SNPs") that reliably represent all of the SNPs in a particular stretch of DNA, tags

that, in effect, promise that a defined block of DNA in which they are located is also present without alteration. Although it will be a long time before the rich variation in the human genome is fully understood, the
HapMap has provided a tool to allow scientists to investigate links between
DNA variation and disease by asking about several hundred thousand
DNA letters instead of sequencing 3 billion, an undertaking that currently
costs 1000 times less than sequencing a full genome.

Known variously as gene chips, DNA chips, and DNA microarrays,
SNP reading technology is basically a collection of thousands of microscopic DNA spots anchored in a precisely ordered way on a solid surface
such as quartz or silicon by a covalent chemical bond. Each short sequence
of DNA, often called a probe, is designed to identify a particular sequence
among the myriad of sequences of a genome under investigation. This detection tool, which permits a *massively parallel* search for SNPs in each sample, is highly informative because of the extreme fidelity of DNA sequences
when they pair. Even a single-base-pair difference will prevent the chemical
connection that would generate a positive signal for the photodetection device used to scan the array after a sample has been layered on it.

Since its founding in 1991, Affymetrix, a California company driven by
the scientific vision of Dr. Stephen Fodor, has led the way in the development of ever more sophisticated DNA microarrays. The first Affymetrix
product, a gene chip to identify variations in the DNA of human immunodeficiency virus (HIV), was introduced in 1994. In 2007, Affymetrix had
(using photolithography) enhanced its fabrication of DNA microarrays to
the point where it could provide the Genome Wide Human SNP Array 6.0,
which features 1.8 million genetic markers, including 906,600 SNP detection cells. Affymetrrix has developed a straightforward method for digest-
ing (chopping into thousands of fragments) the DNA of interest, amplifying the fragments, and then hybridizing them to the probe field on the
array. Hybridization with a probe generates a light signal that is read by a
camera.

For many research projects, this microarray offers far more detection
power than is needed. Affymetrix also makes smaller chips that focus on
particular genes of interest. Although DNA microarrays have thus far been
almost exclusively used in the research setting, the time is not far off when
they will be routinely used as diagnostic tools in the clinical arena. For example, the giant Swiss company, Roche, recently created the Amplichip, an

array that focuses on gene variants that affect how people metabolize drugs and which can help identify patients who may be at increased risk for side effects and/or need a nonstandard dose of a medicine.

By 2006, the tools of SNP analysis made it feasible to interrogate genomes in detail at a cost that created exciting new opportunities in our ever-expanding effort to understand the subtle role that gene variants play in common diseases that have long been known to be more common in some families than in others, but the onset of which was also clearly affected by unknown environmental factors.

Research projects of this nature are called whole-genome (or genome-wide) association studies. The researchers use brute force SNP analysis in an effort to tease out one or more gene variants that play a small but real role in altering risk for diseases such as autism, amyotrophic lateral sclerosis (Lou Gehrig's disease), rheumatoid arthritis, multiple sclerosis, diabetes, and coronary artery disease. Their hope is that by identifying gene variants that confer risk, they will better understand the pathophysiology and gain new insight into possible targets upon which to focus new drug development. Given how intractable diseases such as most of those listed above have been to therapy, it is understandable that whole-genome studies that uncover new ways of thinking about causes of diseases are generating great optimism. Let me provide an example.

In August 2007, the International Multiple Sclerosis Genetics Consortium published its findings in the *New England Journal of Medicine*. An army of researchers assembled 931 family trios (two parents and an affected child). After they applied rigorous clinical rules to confirm the diagnosis (just a few erroneous diagnoses can destroy a genetic study), the laboratory typed 334,923 SNPs in each person. The bioinformatics people then used a statistical test to ask if any SNPs appeared more commonly than would be expected among persons with the disease. If that were true, it would suggest that such SNPs were located in or near a gene variant that conferred risk. The first study found 49 SNPs that were associated with having the disease. The scientists then compared the DNA of the 931 patients with the DNA of 2431 control subjects and found 32 more SNPs that were strongly associated ($P < .001$) with having the disease. They then selected another 40 SNPs that were somewhat less associated ($P < .01$), for a total of 121 SNPs. In what has become a mandatory step in this kind of research, they then used this new panel to investigate a new group of 609 trios. In these subjects, they found two SNPs within a gene called inter-leukin-2 receptor α (IL-2Rα), one SNP within the interleukin-7 receptor α (IL-7Rα) gene, and many within the HLA locus on chromosome 6 that were very strongly associated with having multiple sclerosis. The findings are intellectually gratifying because the proteins encoded by the IL-2Rα and IL-7Rα genes both play an important role in the function of T cells, cells that are central to body's immunological defenses. As multiple sclerosis has long been recognized as an autoimmune disorder, the identification of genes that affect T cells reinforces the central hypothesis. It also provides new ideas for drug development.

Currently, there is a gold rush for whole-genome association studies. In 2007, the nation's leading scientific and medical journals published whole-genome association studies identifying two genes that predispose to psoriasis, several that are associated with the risk of schizophrenia, several that increase the risk of toxicity from an anticancer drug called Cisplatin, many gene variants that are associated with an increased risk for coronary artery disease, one that confers a significant risk for macular degeneration (MD), ten loci that are associated with a modestly increased risk for amyotrophic lateral sclerosis, a very surprising study that showed the close genetic connection between rheumatoid arthritis and lupus, and one that found a variant associated with the pigment disorder known as vitiligo.

The results of the search for genetic variants that increase risk for MD provide an excellent example of the important dividends that genetic studies generate for clinical medicine. MD, the most common cause of blindness in people over 50, is caused by thinning, atrophy, and sometimes bleeding in the central area of the retina known as the macula. This atrophy seems to be caused by deposits called drusen that are related to elevated levels of cholesterol. The drusen damage the retinal cells. Although the precise cause of age-related MD remains elusive, it is well known that age, race (the disorder is more common among whites), family history, cigarette smoking, high blood pressure, a high-fat diet, and high cholesterol levels are risk factors. The lifetime risk for MD is 50% for individuals with an affected first-degree relative compared to only 12% for those with no affected relatives.

In 2005, a team of scientists at Southwestern Medical School in Dallas, building on earlier studies that pointed to a region on chromosome 1 that harbored a gene with variants that predisposed to MD, decided to use

DNA microarrays to see if they could capture the culprit. They tested a large number of SNPs in the suspect region in two separate populations of patients and control groups and found that the correlation between one SNP and risk for the disease was extraordinarily high. The SNP in question actually changed the amino acid structure of a protein called Complement factor H. Statistical analysis showed that people with the SNP had nearly a three-fold higher risk for developing MD than those who did not have it. Further analysis suggested that this common SNP was responsible for a substantial fraction of the risk for MD in the entire population, probably through its effect on the structure of the factor H protein.

In 2006, a group at Duke University Medical Center found that a variant in another gene substantially increased the risk for MD among cigarette smokers, the first demonstration that a common genetic risk factor coupled with a modifiable lifestyle factor greatly increased the risk. In 2007, a research group in Cambridge, England, found yet another relatively common gene variant—this time in the gene for a protein called Complement 3—that also increased risk for MD.

Researchers who report exciting new discoveries about the role of gene variants in risk for serious disorders understandably tend to warn against moving too quickly to use them in the clinic. In the case of MD, however, I believe that one can in 2008 make a persuasive case that DNA risk assessment testing should be made available to the general public. Why? We now understand that because variations in just a couple of genes play a big role in the risk, it would be easy to develop a low-cost test to assess these variations (a test that could be done on cells taken with a cheek swab), and the information would be valuable in counseling about risk and changing behavior to ameliorate it. Of course, no one concerned about good health should smoke. But, if current smokers learned that they carry a gene that greatly increases the risk for blindness among smokers, such information might stimulate them to kick the habit. Among nonsmokers, the discovery at a comparatively early age of a genetic predisposition could drive one to adopt dietary habits (low-fat, high-lutein) that some researchers believe reduce the risk, to be screened more frequently for other risk factors such as cholesterol level and high blood pressure, and to have more frequent eye exams after age 60. In families at high risk, the DNA test could be used to identify those who carry the risk factor and those who do not, thus identifying family members who need special attention and reassuring those who do not that their risk is the same as that found among the general population.

For the next few years, we can foresee that each month some scientific consortium will publish a large study in a group of patients with a common chronic disorder that indicts a new gene. Whole-genome association studies are generating a lot of hope, but they must be interpreted with caution. The studies depend critically on the ability of research physicians to diagnose the disease. But for many of these intractable disorders such as autism, amyotrophic lateral sclerosis, and multiple sclerosis, physicians cannot yet be sure that they are dealing with one disorder or several quite similar disorders. Second, the sheer size of the patient cohort needed for a successful whole-genome study demands millions of dollars, much time, and impressive organizational skills. Third, these studies generate massive amounts of data that do not yield easily to mathematical analysis. Fourth, a whole-genome study can miss a genetic predisposition in a subgroup of the larger cohort. Finally, even the most successful study only finds a statistically suspicious location. The challenge to find which gene in that region of the genome contains the variant that actually is responsible for the association is formidable.

The DNA microarray technology that has spawned whole-genome association research has also provided the technology base for a new consumer-oriented industry that has been dubbed personal genomics. This enterprise is founded on the premise that individuals, who can at relatively low cost now acquire an immense amount of DNA information about themselves, might wish to do so. Although it is not yet generating any revenues, the new industry is generating a lot of buzz, in no small part because one of the first players is a California company called 23andMe (humans have 23 pairs of chromosomes), of which the cofounder is the spouse of a founder of Google. Not surprisingly, the fact that Google made an initial investment in 23andMe caught the attention of the *New York Times*, the *Wall Street Journal*, and many other papers.

23andMe offers customers access to state-of-the-art SNP analysis. It uses a platform made by the Illumina company called the Illumina HumanHap550+ BeadChip, which tests for 550,000 SNPs. In addition, 23andMe has compiled a further set of 30,000 SNPs that it determined to be especially relevant to the suite of services it provides. The well-funded Web-based company has enlisted a troop of prominent physicians to comment on the role of gene variants in common disorders. The business plan is simple: A customer provides a sample (simply by spitting into a container) and the company then performs the analysis, reports it out as a series of risk assessments keyed to the findings in the most current scientific literature, and promises to act as an ongoing source of information as the database grows. It is not clear from the Web site, but it appears that 23andMe outsources the samples for testing. Navigenics, another new California company, offers a similar service, but it is operating its own Clinical Laboratory Improvements Act (CLIA)-approved testing laboratory.

In Cambridge, Massachusetts, another new company is upping the ante. Knome, which bills itself as a "personal genomics company," announced in November of 2007 that it was willing to analyze the entire DNA sequence of a customer for $350,000. If you can afford it, the price seems fair. It took more than a decade and about $3,000,000,000 to compile the first human sequence, a task completed for the most part by 2001. Since then, James Watson (who won the Nobel Prize for codiscovering the shape of the double helix) and Craig Venter (who led a private sequencing effort) are the only two persons on the planet to have had their genomes sequenced. Because the existence of a consensus sequence greatly eases the job of sequencing a new genome, the price for so doing has fallen by a factor of 10,000 in six years! Not surprisingly, Knome, which was cofounded by Harvard Professor George Church, takes a deprecating view of Illumina's array technology, pointing out that it identifies less than 0.02% of the sequence.

A few miles away from Knome, Bill Efcavitch, Senior Vice President of Helicos Biosciences, is betting he can beat Professor Church at his own game. He and his team of "rock star" organic chemists are developing a new kind of sequencing machine. Efcavitch predicts that the new machine will be on the market in just a few years at a cost of about $1.3 million and that it will be capable of sequencing an entire human genome in two weeks at a cost of $10,000. So far, the stock analysts have not fallen in love with Helicos, but that may be because they have not yet grasped the potential size of the personal genomics market.

In 2008, it appears that there will be a clutch of new direct-to-consumer genetic testing companies that focus on risk for psychiatric illness and/or adverse reactions to choice of antidepressant medications. First off the mark is Psynomics, a company built around discoveries made at the University of California (San Diego) concerning the increased risk that certain gene variants may confer for bipolar disease. The decision to launch this test will generate intense controversy. Influential geneticists will almost certainly attack the test as insufficiently researched, of extremely limited predictive value, and potentially dangerous. Psynomics counters that it will make every effort to create a situation in which those (presumably, people with a positive family history) who do take the test receive competent counseling and support.

Of what value are the services offered by these companies? I am very impressed by the amount of personal genetic data that 23andMe and Navigenics will provide given the cost of the service. At first, the vast majority of the data will be of curiosity value, but there will be some bits that will to my mind more than justify the cost (about $1000, which is unlikely to be covered by insurers).

Consider Alzheimer's disease. For more than a decade, it has been well-established that about 2–3% of whites of northern European extraction carry two copies of a gene variant called ApoE4, which codes for a protein involved in lipid transport. Such people are also at much increased risk for developing Alzheimer's disease at age 70 than are similar people who were born with copies of E2 and/or E3. For a long time, the academic medical establishment has clung to the view that an Alzheimer's risk assessment test should not be offered to most individuals (the exception being when a person shows early signs and symptoms of the disorder and the test could help confirm suspicions). Most neurologists thought that because many people with two copies of ApoE4 never develop the disease and many people without two copies do develop it that the test lacked appropriate sensitivity and specificity. I think the test is—at least in some instances—valuable. For example, for individuals with a parent who had two copies of ApoE4 and died with Alzheimer's, it may be reassuring to learn that they have only one copy of E4. Persons with a significant family history of the disease who also have two copies of E4 may decide that they would like to plan their older years in part around the risk of Alzheimer's disease. This may mean planning for earlier retirement, deciding to join a community that offers long-term care, and paying more attention to estate planning, as well as a host of other actions.

Why will most of the other data be of little value? The main reason is that relative risk is very different from absolute risk. If a young woman learns that she has a gene variant that doubles her lifetime risk for multiple sclerosis, it means that her absolute risk may have risen from 1 in 200 to 1 in 100. Because there is little that one can do in response to learning this information and because the absolute risk remains very low, few people will view the data as helpful. Because it would be almost impossible to comment about expected age of onset of a particular disorder, the information will be of no clinical relevance in the near term.

As for Knome's offer to sequence for $350,000, I imagine this will capture the interest of enough hedge fund managers to keep its laboratory busy. When you already own a Gulfstream jet and a 120-foot yacht, paying $350,000 to have your genome sequenced is probably worth the social capital it creates to expend at cocktail parties.

The question that I have most often been asked by persons who are skeptical about the value of genetic information is "So what?" What they mean is that if there is no easy intervention that effectively ameliorates the risk, they do not see a value proposition. Indeed, some perceive that the information would serve only to provoke anxiety. Fair enough. However, as in the case I briefly made for making use of nondiagnostic information about the risk for Alzheimer's disease, there will be a steady increase in the number of disorders for which risk information can be used in a way to reduce risk.

Although in its infancy, personal genomics is already causing substantial controversy. On the one side, scientists, physicians, regulators, and bioethicists are arguing that the services which a rapidly growing number of companies plan to offer are of little value to the consumer and that there is a chance that the information will cause harm. On the other side are business people and some consumers who rally around the principle that individuals should have the right to ask questions about their genome if they wish and who are willing to pay for the service. In addition, proponents argue that market forces will quickly favor good companies with appropriate methods for helping their customers understand the information they purchase.

The debate will probably lead to a showdown that will play out in the courts. In the United States, a federal law known as the Clinical Laboratory Improvements Act requires that labs offering diagnostic tests be certified. In addition, a few states (notably, New York) require an even higher level of regulatory oversight. Most of these regulatory schemes are built around a now long-established premise that only doctors and a few other health care

providers should be permitted to order medical tests. The personalized genomics business is squarely premised on selling directly to consumers, a direct challenge to the standard regulatory model. To circumvent this, some
of the new companies are asserting that the information they generate is
not diagnostic of anything and that their laboratories should not be subject
to limitations such as requiring a physician's order before processing a test
request.

The era of personal genomics is here. The cost of acquiring the information is going to continue to fall and the value of having it is going to
continue to rise. The task of understanding it and figuring out how to
make use of it will take a long time. Still, I am confident that sometime in
the next two decades we will begin to routinely perform genomic analysis
on all newborns, and that salient results will be incorporated into their
medical records to help them chart healthier lives.

Bibliography

CHAPTER 1

THE STRONGEST BOY IN THE WORLD

Grobet L., Martin L.J., Poncelet D., Pirottin D., Bouwers B., Riquet J., et al. 1997. A deletion in the bovine myostatin gene causes the double-muscled phenotype in cattle. *Nat. Genet.* **17:** 4–5.

Holden C. 2004. Peering under the hood of Africa's runners. *Science* **305:** 637–639.

Moran C.N., Scott R.A., Adams S.M., Warrington S.J., Jobling M.A., Wilson R.H., et al. 2004. Y chromosome haplotypes of elite Ethiopian endurance runners. *Hum. Genet.* **115:** 492–497.

Rankinen T., Perusse L., Rauramaa R., Rivera M.A., Wolfarth B., and Bouchard C. 2004. The human gene map for performance and health-related fitness phenotypes: The 2003 update. *Med. Sci. Sports Exerc.* **36:** 1451–1469.

Schuelke M., Wagner K.R., Stolz L.E., Hübner C., Riebel T., et al. 2004. Myostatin mutation associated with gross muscle hypertrophy in a child. *N. Engl. J. Med.* **350:** 2682–2688.

Vogel G. 2004. A race to the starting line. *Science* **305:** 625–632.

Vollind Z.M., Xenophontos S.L., Cariolou M.A., Mokone G.G., Hudson D.E., Anastiasiades L., and Noakes T.D. 2004. The ACE gene and endurance performance during the South African ironman triathlons. *Med. Sci. Sports Exerc.* **36:** 1314–1320.

Willard H. July 14, 2005. The new high-tech sports cheat. *The Boston Globe*, p. A13.

Yang N., MacArthur D.G., Gulbin J.P., Hahn A.G., Beggs A.H., Easteal S., and North K. 2003. ACTN3 genotype is associated with human elite athletic performance. *Am. J. Hum. Genet.* **73:** 627–631.

Zernike K. March 20, 2005. The difference between steroids and Ritalin is.... *The New York Times*, p. WK3.

CHAPTER 2

OUR ANCESTORS

Business Wire, November 15, 2007. Harvard Professor Henry Louis Gates, Jr. Joins Forces with Family Tree DNA to Launch AfricanDNA.com. Online at www.african heritage.com/AfricanDNA_Press_Release.asp.

Dalton R. 2005. Looking for the ancestors. *Nature* **434:** 432–434.

Derenko M., Malyarchuk B., Grzybowski T., Denisova G., Dambueva I., et al. 2007. Phylogeographic analysis of mitochondrial DNA in Northern Asian populations. *Am. J. Hum. Genet.* **81:** 1025–1041.

Forster P. and Matsumura S. 2005. Did early humans go north or south? *Science* **308:** 965–966.

Harmon A. 2007. Stalking strangers' DNA to fill in the family tree, *New York Times*, p. A1.

Johanson D.C. and Edey M.A. *LUCY: The beginnings of humankind.* Simon and Schuster, New York.

Kayser M., Lao O., Kathrin S., Brauer S., Wang X., Nurnberg P., Trent R.J., and Stoneking M. 2008. Genome-wide analysis indicates more Asian than Melanesian ancestry of Polynesians. *Am. J. Hum. Genet.* **82:** 194–198.

Lee F.R. February 5, 2008. Famous black lives through DNA's prism. *New York Times*, p. B1.

Lell J.T., Sukernik R.I., Starikovskaya Y.B., Su B., Jin L., Schurr T.G., et al. 2002. The dual origin and Siberian affinities of Native American Y chromosomes. *Am. J. Hum. Genet.* **70:** 1377–1380.

Macaulay V., Hill C., Achilli A., Rengo C., Clarke D., et al. 2005. Single, rapid coastal settlement of Asia revealed by analysis of complete mitochondrial genomes. *Science* **308:** 1034–1036.

McDougall I., Brown F.H., and Fleagle J.G. 2005. Stratigraphic placement and age of modern humans from Kibish, Ethiopia. *Nature* **433:** 733–736.

Mietto P., Avanzini M., and Rolandi G. 2003. Palaeontology: Human footprints in Pleistocene volcanic ash. (Brief communications.) *Nature* **422:** 133.

Mulligan C., Hunley K., Cole S., and Long J.C. 2004. Population genetics, history, and health patterns in Native Americans. *Annu. Rev. Genomics Hum. Genet.* **5:** 295–315.

Salas A., Richards M., Lareu M.-V., Scozzari R., Coppa A., Torroni A., Macaulay V., and Carracedo A. 2004. The African diaspora: Mitochondrial DNA and the Atlantic slave trade. *Am. J. Hum. Genet.* **74:** 454–464.

Sciolino E. October 11, 2007. Plan to test DNA of some immigrants divides France. *New York Times*, p A13.

Shipman P. 2001. *The man who found the missing link: Eugene Dubois and his lifelong quest to prove Darwin right.* Simon & Schuster, New York.

Wade N. May 13, 2005. DNA study yields clues on early humans' first migration. *The New York Times*, p. A7.

Wells S. 2003. *The journey of man.* Random House, New York.

Wilford J.N. December 13, 1996. 3 human species coexisted on earth, new data suggest. *The New York Times*, p. A1.

Winstein K.J. November 15, 2007. Harvard's Gates refines genetic-ancestry searches for blacks. *Wall Street Journal*, p. D5.

Zerjal T., Xue Y., Bertorelle G., Wells R.S., Bao W., Zhu S., Qamar R., et al. 2003. The genetic legacy of the Mongols. *Am. J. Hum. Genet.* **72:** 717–721.

CHAPTER 3

RACE

Bloche M.G. 2004. Race based therapeutics. *N. Engl. J. Med.* **351:** 2035–2036.

Burchard E.G., Ziv E., Coyle N., Gomez S.L., Tang H., Karter A.J., et al. 2003. The importance of race and ethnic background in biomedical research and clinical practice. *N. Engl. J. Med.* **248:** 1070–1075.

Cavalli-Sforza L.L. 2000. *Genes, peoples, and languages.* Farrar, Strauss and Giroux, New York.

Dobzhansky T. 1966. *Heredity and nature of man.* New American Library, New York.

Dunn L.C. and Dobzhansky T. 1952. *Heredity, race and society.* New American Library, New York.

Genetics for the Human Race. 2004. *Nat. Genet.* (suppl.) **36:** S1–S60. (Twelve articles on genetics and race.)

Harmon A. July 25, 2005. Blacks pin hope on DNA to fill slavery's gaps in family trees. *The New York Times,* p. 1.

Stanton W.R. 1960. *The leopard's spots: Scientific attitudes toward race in America 1815–59.* University of Chicago Press, Illinois.

Sykes B. 2001. *The Seven Daughters of Eve.* W.W. Norton & Company, New York.

CHAPTER 4

LONGEVITY

Brenner S. Curriculum vitae. Online at nobelprize.org/medicine/laureates/ 2002 /brenner-cv.html.

Brenner S. 1974. The genetics of *Caenorhabditis elegans. Genetics* **77:** 71–94.

Carey J.R. 2003. *Longevity: The biology and demography of life span.* Princeton University Press, New Jersey.

Hamet P. and Tremblay J. 2003. Genes of aging. *Metabolism* (suppl. 2) **52:** 525–529.

Helfand S.L. and Blanka R. 2003. Genetics of aging in the fruit fly, *Drosophila melanogaster. Annu. Rev. Genet.* **37:** 329–348.

Kin K., Honor H., Libina N., and Kenyon C. 2001. Regulation of the *Caenorhabditis elegans* longevity protein DAF-16 by insulin/IGF-1 and germline signaling. *Nat. Genet.* **28:** 139–145.

Murakami S. and Johnson T.E. 2003. Regulation of life span in model organisms. *Curr. Genomics* **4:** 63–74.

Perls T. 1995. The oldest old. *Sci. Am.* **272:** 70–75. Online at www.sciam.com.

Scanlon L. 2004. The longevity gene. *MIT News.* Online at www.techreview.com/ articles/o4/10/scanlon1004.asp?p=0.

Zaslow J. March 12, 2005. This club has one requirement: 110 birthdays. *The Wall Street Journal,* p. A1.

CHAPTER 5

INTELLIGENCE

Benbow C.P., Lubinski D., Shea D.L., and Eftekhari-Sanjani H. 2000. Sex differences in mathematical reasoning ability at age 13: Their status 20 years later. *Psychol. Sci.* **11:** 474–480.

Devlin B., Daniels M., and Roeder K. 1997. The heritability of IQ. *Nature* **388:** 468–471.

Gould S.G. 1981. *The mismeasure of man.* Norton, New York.

Herrnstein R.J. and Murray C. 1996. *The bell curve: Intelligence and class structure in American life.* Simon & Schuster, New York.

Hyde J.S., Fennema E., and Lamon S.J. 1990. Gender differences in mathematics performance: A meta-analysis. *Psychol. Bull.* **107:** 139–155.

Plomin R., Defries J.C., Craig I.W., and McGuffin P., eds. 2003. *Behavioral genetics in the postgenomic era.* American Psychological Association, Washington D.C.

Plomin R., Hill L., Craig I.W., McGuffin P., Purcell S., Sham P., et al. 2001. A genome-wide scan of 1842 markers for allelic associations with general cognitive ability: A five-stage design using DNA pooling and extreme selected groups. *Behav. Genet.* **31:** 409–497.

Posthuma D., Luciano M., de Geus E.J.C., Wright M.J., Slagboom P.E., et al. 2005. A genomewide scan for intelligence identifies quantitative trait loci on 2q and 6p. *Am. J. Hum. Genet.* **77:** 318–326.

Terman L.M. and Oden M.H. 1947. *Genetic studies of genius.* Volume IV. *The gifted child grows up.* Stanford University Press, California.

Whalen J. and Begley S. March 30, 2005. In England, girls are closing gap with boys in math. *The Wall Street Journal,* p. A1.

CHAPTER 6

CHARCOT-MARIE-TOOTH DISEASE

Bertorini T., Narayanaswami P., and Rashed H. 2004. Charcot-Marie-Tooth disease (hereditary motor sensory neuropathies) and hereditary sensory and autonomic neuropathies. *Neurologist* **10:** 327–337.

Bird T.D. 1999. Historical perspective of defining Charcot-Marie-Tooth type 1B. *Ann. N.Y. Acad. Sci.* **883:** 6–13.

Brody I.A. and Wilkens R.H. 1967. Charcot-Marie-Tooth disease. *Arch. Neurol.* **17:** 552–557.

Grandis M. and Shy M.E. 2005. Current therapy for Charcot-Marie-Tooth disease. *Curr. Treat. Options Neurol.* **7:** 223–231.

Houlden H., Blake J., and Reilly M.M. 2004. Hereditary sensory neuropathies. *Curr. Opin. Neurol.* **17:** 569–577.

Pareyson D. 2003. Diagnosis of hereditary neuropathies in adult patients. *J. Neurol.* **250:** 148–160.

Pearce J.M. 2000. Howard Henry Tooth (1856–1925). *J. Neurol.* **247:** 3–4.

Smith A.G. 2001. Charcot-Marie-Tooth disease. *Arch. Neurol.* **58:** 1014–1016.

CHAPTER 7

HUNTINGTON'S DISEASE

Codori A.-M., Slavney P.R., Rosenblatt A., and Brandt J. 2004. Prevalence of major depression one year after predictive testing for Huntington's disease. *Genet. Test.* **8:** 114–118.

Couzin J. 2004. Huntington's disease: Unorthodox clinical trials meld science and care. *Science* **304:** 816–817.

DeJong R.N. 1973. The history of Huntington's chorea in the United States of America. *Adv. Neurol.* **1:** 19–27.

Gusella J.F., Wexler N.S., Conneally P.M., Naylor S.L., Anderson M.A., Tranzi R.E., Watkins P.H., Ottina K., Wallace M.R., Sakaguchi A.V., Young A.M., Shoulson I., Bonilla E., and Martin J.B. 1983. A polymorphic DNA marker genetically linked to Huntington's disease. *Nature* **306:** 234–238.

Huntington's Disease Collaborative Research Group. 1993. A novel gene containing a trinucleotide repeat that is expanded and unstable on Huntington's disease chromosomes. *Cell* **72:** 971–983.

Lee S.T., Chu K., Park J.E., Lee K., Kang L., Kim S.U., and Kim M. 2005. Intravenous administration of human neural stem cells induces functional recovery in Huntington's disease rat model. *Neurosci. Res.* **52:** 243–249.

Melone M.A., Jor F.P., and Peluso G. 2005. Huntington's disease: New frontiers for molecular and cell therapy. *Curr. Drug Targets* **6:** 43–56.

Ryu R.K., Kim J., Cho S.J., Hatori K., Nagai A., Choi H.B., Lee M.C., McLarnon J.G., and Kim S.U. 2004. Proactive transplantation of human neural stem cells prevents degeneration of striatal neurons in a rat model of Huntington disease. *Neurobiol. Discov.* **16:** 6877.

CHAPTER 8

DEAFNESS

Davey M. March 21, 2005. As town for the deaf takes shape, debate on isolation re-emerges. *The New York Times*, p. A1.

Davies P. March 29, 2005. Toddlers' implants bring upheaval to deaf education. *The Wall Street Journal*, p. A1.

Dennis C. 2004. Genetics: Deaf by design. *Nature* **431:** 894–896. (News feature.)

Hladek G.A. 2002. Cochlear implants, the deaf culture, and ethics: A study of disability, informed surrogate consent, and ethnocide. *Monash Bioeth. Rev.* **21:** 29–44.

Hudspeth A.J. 1989. How the ear's works work. *Nature* **341:** 397–404.

Nance W.E. 2003. The genetics of deafness. *Ment. Retard. Dev. Disabil. Res. Rev.* **9:** 109–119.

Nance W.E., Liu X.Z., and Pandya A. 2000. Relation between choice of partner and high frequency of connexin-26 deafness. *Lancet* **356:** 500–501.

Stern R.E., Yueh B., Lewis C., Norton S., and Sie K.C. 2005. Recent epidemiology of pediatric cochlear implants in the United States: Disparity among children of different ethnicity and socioeconomic status. *Laryngoscope* **115:** 125–131.

Wade N. February 1, 2005. A new language arises, and scientists watch it evolve. *The New York Times*, p. D3.

Zeng F.G. 2004. Trends in cochlear implants. *Trends Amplif.* **8:** 1–34.

CHAPTER 9

SAN LUIS VALLEY SYNDROME

Becher M.W., Morrison L., Davis L.E., Maki W.C., King M.K., Bicknell J.M., et al. 2001. Oculopharyngeal muscular dystrophy in Hispanic New Mexicans. *J. Am. Med. Assoc.* **286:** 2437–2440.

Graw S.L., Sample T., Bleskan J., Sujansky E., and Patterson D. 2000. Cloning, sequencing, and analysis of Inv8 chromosome breakpoints associated with recombinant 8 syndrome. *Am. J. Hum. Genet.* **66:** 1138–1144.

Sujansky E., Smith A.C., Peakman D.C., McConnell T.S., Baca P., and Robinson A. 1981. Familial pericentric inversion of chromosome 8. *Am. J. Med. Genet.* **10:** 229–335.

Sujansky E., Smith A.C., Prescott K.E., Freehauf C.L., Clericuszio C., and Robinson A. 1993. Natural history of the recombinant (8) syndrome. *Am. J. Med. Genet.* **47:** 512–525.

Yi Z., Garrsion N., Cohen-Barak O., Karafet T.M., King R.A., Erickson R.P., et al. 2003. A 122.5-kilobase deletion of the P gene underlies the high prevalence of oculocutaneous albinism type 2 in the Navajo population. *Am. J. Hum. Genet.* **72:** 62–72.

CHAPTER 10

SEVERE COMBINED IMMUNE DEFICIENCY

Berns A. 2004. Good news for gene therapy. *N. Engl. J. Med.* **350:** 1679–1680.

Check E. 2002. Gene therapy: A tragic setback. *Nature* **420:** 116–118. (News feature.)

Hacein-Bey-Albina S., Le Driest F., Carlier F., Bouneaud C., Hue C., De Villartay J.P., et al. 2002. Sustained correction of X-linked severe combined immune deficiency by ex vivo gene therapy. *N. Engl. J. Med.* **346:** 1185–1193.

Marshall E. 2003. Gene therapy. Second child in French trial is found to have leukemia. *Science* **299:** 320.

McCormack M.P. and Rabbitts T.H. 2004. Activation of the T-cell oncogene LMO2 after gene therapy for X-linked severe combined immunodeficiency. *N. Engl. J. Med.* **350:** 913–922.

National Marrow Donor Program. Online at www.marrow.org.

Noguchi P. 2003. Risks and benefits of gene therapy. *N. Engl. J. Med.* **348:** 193–194.

CHAPTER 11

DOGS

Cargill E.J., Famula T.R., Strain G.M., and Murphy K.E. 2004. Heritability and segregation analysis of deafness in U.S. dalmatians. *Genetics* **166:** 1385–1393.

Hare B., Brown M., Williamson C., and Toamsello M. 2002. The domestication of social cognition in dogs. *Science* **298:** 1634–1636.

k9 Genetics Corp. Online at www.k9genetics.com.

Kaminski J., Call J., and Fischer J. 2004. Word learning in a domestic dog: Evidence for "fast mapping." *Science* **304:** 1682–1683.

Kolata G. August 4, 2005. Beating hurdles, scientists clone a dog for a first. *The New York Times*, p. 1.

Leonard J.A., Wayne R.K., Wheeler J., Valadez R., Guillen S., and Vila C. 2002. Ancient DNA evidence for Old World origin of New World dogs. *Science* **298:** 1613–1616.

Lohi H., Young E.J., Fitzmaurice S.N., Rusbridge C., Chan E.M., Vervoor L.M., et al. 2005. Expanded repeat in canine epilepsy. *Science* **307:** 81.

Ostrander E. and Giniger E. 1997. Semper fidelis: What man's best friend can teach us about human biology and disease. *Am. J. Hum. Genet.* **61:** 475–480.

Parker H.G., Kim L.V., Sutter N.B., Carlson S., Lorentzen T.D., Malek T.B., et al. 2004. Genetic structure of the purebred domestic dog. *Science* **304:** 1160–1164.

Pennisi E. 2004. Genome resources to boost canines' role in gene hunts. *Science* **304:** 1093–1094.

Spady T.C. and Ostrander E.A. 2008. Canine behavioral genetics: Pointing out the phenotypes and herding up the genes. *Am. J. Hum. Genet.* **82:** 10–18.

Van Hagen M.A., Janss L.L., van den Brock J., and Knol B.W. 2004. The use of a genetic-counseling program by Dutch breeders for four hereditary health problems in boxer dogs. *Prev. Vet. Med.* **63:** 39.

Vogan K. December 2007. Canine nutrigenomics. *Nat. Genet.* **39:** 1427.

CHAPTER 12

CATS

The Cat Genome. Online at lgd.abcc.ncicrf.gov.

Darnton R. 1985. *The great cat massacre and other episodes in French cultural history.* Vintage Press, New York.

Eisenberg A. May 28, 2005. Hello kitty, hello clone. *The New York Times*, p. B5.

O'Brien S.J. 2003. *The tears of the cheetah.* St. Martin's Griffin, New York.

O'Brien S.J., Menotti-Raymond M., Murphy W.J., and Yuhki N. 2002. The feline genome project. *Annu. Rev. Genet.* **36:** 657–686.

Pontius J.U., Mullikin J.C., Smith D.R., Agencourt Sequencing Team, Lindblad-Tor K., et al. 2007. Initial sequence and comparative analysis of the cat genome. *Genome Res.* **17:** 1547–1549.

Uphyrkina O., Miquelle D., Quigley H., Driscoll C., and O'Brien A.J. 2002. Conservation genetics of the far eastern leopard (*Panthera pardus orientalis*). *J. Hered.* **93:** 303–311.

Vigne J.-D., Guilaine J., Debue K., Haye L., and Gerard P. 2004. Early taming of the cat in Cyprus. *Science* **304:** 259.

Warner M. April 29, 2002. Inside the very strange world of billionaire John Sperling. *Fortune*, pp. 99–102.

Weiss J. February 15, 2002. Cloned cat raises prospect of boutique pet creation. *Boston Globe*, p. A3.

CHAPTER 13

MICE

Culliton B.J. 1972. The Jackson laboratory: "Mice are our most important product." *Science* **177:** 871–874.

Guenet J.L. and Bonhomme F. 2003. Wild mice: An ever-increasing contribution to a popular mammalian model. *Trends Genet.* **19:** 24–31.

The Jackson Laboratory. Online at www.jax.org.

Little C.C. 1924. The genetics of tissue transplantation in mammals. *J. Cancer Res.* **8:** 75–95.

Little C.C. and McPheters B.W. 1932. The incidence of mammary cancer in a cross between two strains of mice. *Am. Nat.* **66:** 568–571.

Nagy A., Gertenstein M., Vintersten K., and Behringer R. 2003. *Manipulating the mouse embryo: A laboratory manual*, 3rd edition. Cold Spring Harbor Laboratory Press, Cold Spring Harbor, New York.

Pennacchio L.A. 2003. Insights from human/mouse genome comparisons. *Mamm. Genome* **7:** 429–436.

Rader K. 2004. *Making mice: Standardizing animals for American biomedical research, 1900–1955.* Princeton University Press, New Jersey.

Wade N. September 7, 1999. Of smart mice and an even smarter man. *The New York Times*, p. A1.

Waterson R.H., Lindblad-Toh K., Birney E., Rogers J., Abril J.F., et al. 2002. Initial sequencing and comparative analysis of the mouse genome. *Nature* **420:** 520–562.

Williams R.W., Flaherty L., and Threadgill D.W. 2003. The math of making mutant mice. *Genes Brain Behav.* **4:** 191–200.

CHAPTER 14

CORN

Cohen J.I. 2005. Poorer nations turn to publicly developed GM crops. *Nat. Biotechnol.* **23:** 27–33.

Federoff N.V. 2003. Agriculture. Prehistoric GM corn. *Science* **302:** 1158–1159.

Grove-White R., Macnaghten P., Mayer S., and Wynne B. 1997. *Uncertain world: Genetically modified organisms, food and public attitudes in Britain.* Center for the Study of Environmental Change. Lancaster University, United Kingdom.

Harjes C.E., Rocheford T.R., Bai L., Brutnell T.P., Kandianis C.B., et al. 2008. Natural genetic variation in lycopene epsilon cyclase tapped for maize biofortification. *Science* **310:** 330–333.

Hill W.G. 2005. A century of corn selection. *Science* **307:** 683–684.

Keller E.F. 1983. *A feeling for the organism: The life and work of Barbara McClintock.* W.H. Freeman, New York.

MacNeish R.S. 1964. The origins of new world civilization. *Sci. Am.* **211:** 29–37.

Mangelsdorf P.C., MacNeish R.S., and Galinat W.C. 1964. Domestication of corn. *Science* **143:** 538–545.

Pollack A. July 28, 2004. Panel sees no unique risk from genetic engineering. *The New York Times.*

Quist D. and Chapela I.H. 2001. Transgenic DNA introgressed into traditional maize land races in Oaxaca, Mexico. *Nature* **414:** 541–543.

Specter M. April 10, 2000. The Pharmageddon riddle: Did Monsanto just want more profits or did it want to save the world? *The New Yorker*, pp. 58–71.

Wright S.I., Bi I.V., Schroeder S.G., Yamasaki M., Doebley J.F., McMullern M.D., and Gaut B.S. 2005. The effects of artificial selection on the maize genome. *Science* **308:** 1310–1314.

CHAPTER 15

RICE

Beyer P., Al-Babili S., Ye X., Lucca P., Schaub O., Welsch R., and Potrykus I. 2002. Golden rice: Introducing the beta-carotene biosynthesis pathway into rice endosperm by genetic engineering to defeat vitamin A deficiency. *J. Nutr.* **132:** 506S–510S.

Brody J. January 11, 2005. Facing biotech foods without the fear factor. *The New York Times,* p. D7.

Cyranoski D. 2005. Pesticide results help China edge transgenic rice towards market. *Nature* **435:** 3.

Dugger C.W. October 10, 2007. A bounty of rice seed, out of reach in Africa. *New York Times,* p. A10.

Editorial. 2005. Reburnishing golden rice. *Nat. Biotechnol.* **23:** 395.

Grusak M.A. 2005. Golden rice gets a boost from maize. *Nat. Biotechnol.* **23:** 429–430.

Huang J., Hu R., Rozelle S., and Pray C. 2005. Insect-resistant GM rice in farmers' fields: Assessing productivity and health effects in China. *Science* **308:** 688–690.

Huke R.E. and Huke E.H. 1990. *Rice: Then and now.* International Rice Research Institute, Laguna, Phillippines. Online at www.riceweb.org/History.htm.

International Rice Genome Sequencing Project. 2005. The map-based sequence of the rice genome. *Nature* **436:** 793–800.

Lucca P., Hurrell R., and Potrykus I. 2002. Fighting iron deficiency anemia with iron-rich rice. *J. Am. Coll. Nutr.* **21:** 184S–190S.

Normile D. 2000. Monsanto donates its share of golden rice. *Science* **289:** 843.

Potrykus I. 2004. Experience from the humanitarian golden rice project: Extreme precautionary regulation prevents use of green biotechnology in public projects. *BioVision Alexandria 3–6,* April 2004. Online at www.agbioworld.org/biotech_info/articles/potrykus.html.

Tyagi A.K., Khurana J.P., Khurana P., Raghuvanshi S., Gaur A., et al. 2004. Structural and functional analysis of the rice genome. *J. Genet.* **83:** 79–99.

Welch R.M. and Graham R.D. 2004. Breeding for micronutrients in staple food crops from a human nutrition perspective. *J. Exp. Bot.* **55:** 353–364.

CHAPTER 16

HISTORY

Edwards T. November 23, 1998. Family reunion. *Time,* pp. 87–88.

Gray M.J. March 3, 2001. A founding father and his family ties. *The New York Times,* p. A11.

Krajick K. May 16, 2005. The mummy doctor. *The New Yorker,* pp. 66–75.

Tomsho R. and Nelson E. November 24, 2004. Could DNA tests solve the mystery of Miles Standish? *The Wall Street Journal,* p. A1.

Webster P. December 16, 1999. DNA may solve riddle of French king's death. *The Guardian.* Online at www.guardian.co.uk?Print/o,3858,394231,00.html.

CHAPTER 17

DNA FORENSICS

American Civil Liberties Union, Forensic DNA Databases. Online at www.aclu.org/privacy/biotech.

Belluck P. January 10, 2005. To try to net killer, police ask a small town's men for DNA. *The New York Times*, p. A1.

Boyer P.J. January 17, 2000. Annals of justice: DNA on trial. *The New Yorker*, pp. 42–53.

Dewan S.K. July 19, 2004. Georgia capital case tests the reach of DNA appeals. *The New York Times*, p. A11.

Franklin-Barbajosa C. May 1992. The new science of identity. *National Geographic*, pp. 112–123.

Lazer D., ed. 2004. *DNA and the criminal justice system: The technology of justice.* MIT Press, Cambridge.

Lichtblau E. March 12, 2003. New federal plan for DNA testing is proposed. *The New York Times*, p. A17.

Romero S. November 4, 2004. Outsider claims kinship to Texas ranch dynasty. *The New York Times*, p. A14.

The Innocence Project. Benjamin N. Cardozo School of Law, New York. Online at www.innocenceproject.org.

York M. September 3, 2004. A pilot's fate in Vietnam is discovered through DNA. *The New York Times*, p. A14.

CHAPTER 18

ART AND LANGUAGE

Anker S. and Nelson D. 2004. *The molecular gaze: Art in the genetic age.* Cold Spring Harbor Laboratory Press, Cold Spring Harbor, New York.

Fiedler L. 1978. *Freaks: Myths and images of the secret self.* Simon & Schuster, New York.

James C. April 22, 2005. Duplicates that can beat the real thing. *The New York Times*, p. B29.

Nagourney A. January 13, 2005. Kennedy warns Democrats not be Republican clones. *The New York Times*, p. A23.

Rough B.J. January 30, 2005. His genes have gifts. Mine carry risk. *The New York Times*, p. 7.

Turner J. 1998. *Frankenstein's footsteps: Science, genetics, and popular culture.* Yale University Press, New Haven.

Young M. March 20, 2005. The survivor. *The New York Times*, pp. 30–35.

CHAPTER 19

PREIMPLANTATION GENETIC DIAGNOSIS

Belkin L. July 1, 2001. The made-to-order savior: Producing a perfect baby. *The New York Times Magazine*, pp. 158–169.

Braude P. 2006. Preimplantation diagnosis for genetic susceptibility. *N. Engl. J. Med.* **355:** 541–543.

Kmietowicz Z. 2004. UK clinic allowed to screen embryos for rare bowel cancer. *Br. Med. J.* **329:** 1061.

The Preimplantation Genetic Diagnosis International Society (PGDIS). 2007. Guidelines for good practice in PGD: programme requirements and laboratory quality assurance. *Reproductive Medicine Online* **16:** 134–147. Online at www.rmbonline. com/Article3276.

Sermon K., Van Steriteghem A., and Liebaers I. 2004. Preimplantation genetic diagnosis. *Lancet* **363:** 1633–1641.

Sheldon S. and Wilkinson S. 2004. Should selecting saviour siblings be banned? *J. Med. Ethics* **30:** 533–537.

Thornbill A.R., deDie-Smulders C.E., Geraedts J.P., Harper J.C., Harton G.L., et al. 2004. ESHRE PGD consortium "Best practice guidelines for clinical preimplantation genetic diagnosis (PGD) and preimplantation genetic screening (PGS)." *Hum. Reprod.* Published online November 11, 2004 at humrep.oupjournals.org/ cgi/content/short/deh579v1.

Verlinsky Y., Rechitsky S., Verlinsky O., Ozen S., Beck R., and Kulier A. 2004. Preimplantation genetic diagnosis for polycystic kidney disease. *Fertil. Steril.* **82:** 926–929.

Wolf S.M., Kahn J.P., and Wagner J.E. 2003. Using preimplantation genetic diagnosis to create a stem cell donor: Issues, guidelines and limits. *J. Law Med. Ethics* **31:** 327–339.

CHAPTER 20

STEM CELLS

Edwards B.E., Gearhart J.D., and Wallach E.E. 2000. The human pluripotent stem cell: Impact on medicine and society. *Fertil. Steril.* **74:** 1–7.

Gearhart J. 1998. New potential for human embryonic stem cells. *Science* **282:** 1061–1062.

Greenberger S.S. March 29, 2005. Senate bill sets rules for stem cell research. *Boston Globe,* p. A1.

Hanna J., Wernig M., Markouski S., Sun C.W., Meissner A., Cassady J.P., Beard C., Brambrin T., Wu L.C., Townes T.M., and Jaenisch R. 2007. Treatment of sickle cell anemia mouse model with IPS cells generated from autologous skin. *Science* **318:** 1920–1923.

Hochedlinger K. and Jaenisch R. 2003. Nuclear transplantation, embryonic stem cells, and the potential for cell therapy. *N. Engl. J. Med.* **349:** 275–286.

Hwang W.S., Ryu Y.J., Park J.H., Park E.S., Lee E.G., Koo J.M., et al. 2004. Evidence of a pluripotent human embryonic stem cell line derived from a cloned blastocyst. *Science* **303:** 1669–1674.

Kolat G. November 21, 2007. Scientists bypass need for embryo to get stem cells. *New York Times,* p. A1.

Korblin M. and Estriov Z. 2003. Adult stem cells for tissue repair—A new therapeutic concept? *N. Engl. J. Med.* **349:** 570–582.

Mignone J.L., Kukekov V., Chiang A.S., Steindler D., and Enikolopov G. 2004. Neural stem and progenitor cells in nestin-GFP transgenic mice. *J. Comp. Neurol.* **469:** 311–324.

Orkin S. 2000. Stem cell alchemy. *Nat. Med.* **11:** 1212–1213.

Pollack A. February 17, 2005. Moving stem cells front and center. *The New York Times,* p. C1.

Rosenthal N. 2003. Prometheus's vulture and the stem-cell promise. *N. Engl. J. Med.* **349:** 267–274.

Sandel M.J. 2004. Embryo ethics—The moral logic of stem-cell research. *N. Engl. J. Med.* **351:** 207–211.

Shamblott M.J., Axelman J., Littlefield J.W., Blumenthal P.D., Huggins G.R., Cui Y., et al. 2001. Human embryonic germ cell derivatives express a broad range of developmentally distinct markers and proliferate extensively in vitro. *Proc. Natl. Acad. Sci.* **98:** 113–118.

Solter D. and Gearhart J. 1999. Putting stem cells to work. *Science* **283:** 1468–1470.

Stemagen, Press Release. January 17, 2008. Major advance towards creating patient-specific and disease-specific stem cells for therapeutic use. Online at www.stemagene.com.

Stem Cell Research and Ethics. 2000. *Science* **287:** 1417–1144. (Twelve articles about stem cell research.)

Takahasshi K., Tanabe K., Ohnuki M., Narita M., Ichisaka T., Tomoda K., and Yamanaka S. 2007. Induction of pluripotent stem cells from adult human fibroblasts by defined factors. *Cell* **131:** 861–872.

Thompson J.A., Itskovitz-Eldor J., Shapiro S.S., Waknitz M.A., Swiergiel J.J., Marshall V.S., and Jones J.M. 1998. Embryonic stem cell lines derived from human blastocysts. *Science* **282:** 1145–1147.

Vogel G. 2005. Korean team speeds up creation of cloned human stem cells. *Science* **308:** 1096–1097.

CHAPTER 21

PERSONAL GENOMICS

Bahcall O. 2007. Coverage and power. *Nat. Genet.* **39:** 1429.

Christensen K. and Murray J. 2007. What genome-wide association studies can do for medicine. *N. Engl. J. Med.* **356:** 1094–1097.

Couzin J. 2008. Gene tests for psychiatric risk polarize researchers. *Science* **319:** 274–277.

Edwards A.O., Ritter R., Abel K.J., Manning A., Panhuysen C., and Farrer L.A. 2005. Complement Factor H polymorphism and age-related macular degeneration. *Science* **308:** 421–424.

Hunter D.J., Khoury M.J., and Drazen J.M. 2008. Letting the genome out of the bottle—Will we get our wish? *N. Engl. J. Med.* **358:** 105–106.

The International HapMap Consortium. 2003. The International HapMap Project. *Nature* **426:** 789–796.

The International Multiple Sclerosis Genetics Consortium. 2007. Risk alleles for multiple sclerosis identified by a genomewide study. *N. Engl. J. Med.* **357:** 851–862.

Kolata G. January 15, 2008. $300 to learn risk of cancer of the prostate. *New York Times*, p. A1.

Samani N.J., Erdmann J., Hall A.S., Hengstenberg C., Mangino M., et al. 2007. Genomewide association analysis of coronary artery disease. *N. Engl. J. Med.* **357:** 443–453.

SNPedia. Online at http://www.SNPedia.com.

Spencer K.L., Hauser M.A., Olson L.M., Schnetz-Boutand N., Scott W.K., Schmidt S., Gallins P., Agarwal A., Postel E.A., Pericek-Vance M.A., and Hines J.L. 2007. Haplotypes spanning the complement factor H gene are protective against age-related macular degeneration . *Invest. Ophthalmol. Vis. Sci.* **48:** 4277–4283.

United States Department of Health and Human Services. 2007. Personalized health care: Opportunities, pathways, resources. Online at www.hhs.gov/myhealthcare/.

Wade N. September 4, 2007. In the Genome Race, the sequel is personal. *New York Times*, p. D1.

Winslow R. November 6, 2007. Is there a heart attack in your future? *Wall Street Journal*, p. D1.

Wolfberg A.F. 2006. Genes on the web—Direct-to-consumer marketing of genetic testing. *N. Engl. J. Med.* **355:** 543–545.

Index

293